*In the World of*

# GEOGRAPHY

by
Wilbur Cross

cartoon illustrations by
Tom Kerr

BARRON'S

*All inquiries should be addressed to:*
Barron's Educational Series, Inc.
250 Wireless Boulevard
Hauppauge, New York 11788

Library of Congress Catalog Card No. 90-28924

International Standard Book No. 0-8120-4480-0

**Library of Congress Cataloging-in-Publication Data**
Cross, Wilbur.
    Who, what, when, where, why—in the world of
geography / by Wilbur Cross ; cartoon illustrations
by Tom Kerr.
        p.   cm. — (Barron's whiz quiz series)
    Summary: A collection of questions and answers about
countries, bodies of water, deserts, polar regions,
maps, people, natural disasters and other topics relating
to geography.
    ISBN 0-8120-4480-0
    1. Geography—Miscellanea—Juvenile literature.
[1. Geography—Miscellanea.   2. Questions and
answers.]   I. Title.   II. Series.
G133.C76      1991
910—dc20                                    90-28924
                                                CIP

PRINTED IN THE UNITED STATES OF AMERICA
1 2 3 4   5500   9 8 7 6 5 4 3 2 1

# Contents

# Introduction

Geographers used to be a pretty stuffy bunch of people, spending all their time poring over maps and debating among themselves which borders of which countries needed to be changed and what the valid spellings should be for remote and unpronounceable place names. They communicated little with any of the people whose lands they were compiling information about and tended to speak more to each other than to ordinary folks like you and me. They were quite smug, too, knowing that there would always be work for them to do, since the geography books they compiled today would be outdated tomorrow—and sometimes even by the time they came off press.

But all that changed, particularly when organizations like the National Geographic Society began presenting material that was colorful, people-oriented, and sometimes quite provocative. They even displayed flashes of good humor, as well as personal interest in their readers.

"Have you heard of the lost generation?" recently asked Gilbert M. Grosvenor, president of the National Geographic Society, more with tongue in cheek than a scolding voice. Answering his own rhetorical question, he replied, "We have found them. They are lost. They haven't the faintest idea where they are."

He was referring to a 1988 survey that revealed a curious fact: Despite their spirit of adventure and unprecedented travels to far places, many young Americans really had a rather dismal grasp of geography. In fact, in a test taken by participants from nine countries, Americans came in dead last! The situation has been changing since then, as more and more people have come to realize that geography can be *fun*, as well as essential to our understanding of these rapidly changing times.

Just what *is* geography? We tend to overlook that it relates as much to people, politics, and events as it does to places. After all, what is an atlas but a record of all of the things that have happened, and are happening, to people and civilizations? That includes not only natural cataclysms like volcanic upheavals, earthquakes, floods, and storms, but man-made events such as voyages of discovery, wars, and sweeping changes in governments and the balances of power. Put your finger on any point on the globe and you have a source for fascinating records of history, drama, science, and wonder. Is there a place on earth that does not have a story to tell? And is there any moment when geographical facts about these places are not timely? Through generations of research, and occasionally pure trial and error, geographers are able to reconstruct the way portions of the earth looked thousands and even millions of years ago and to chart what these areas will look like by the end of the next millenium. So we can live yesterday, today, and tomorrow just by looking at maps and reflecting on what they mean.

This book is not intended to be a comprehensive review of our geographical world. That would take the proverbial "five foot shelf of books" to accomplish. What it is, rather, is a smorgasbord of facts. Some fit the "trivia" category, others are fundamentals of geography. The individual questions in this book all relate to one major question: Why and in what manner is any given place on earth different from others and how is it likely to change in nature temporarily or permanently? This is one of the most essential questions of geography and the reason why it has become a science in its own right during thousands of years of study and development.

We hope you will find the text entertaining, the facts provocative, and the questions challenging.

# Acknowledgments

I am indebted to a good many sources for the information compiled and used in these questions and answers, particularly public libraries in the states of New York, New Hampshire, Georgia, and South Carolina, where I happened to be residing for periods of time during the course of researching and writing the book. My list of reference works, articles, and other publications numbers close to forty. I appreciated many suggestions from my wife, four daughters, two sons-in-law, two grandchildren, and a host of friends. I am specially grateful to Judy Makover and Diane Roth for the extensive editing and multitudes of suggestions they made, to Grace Freedson for getting me involved with the assignment to begin with, and to Phillip Lefton, who checked the manuscript for scientific flaws or misleading questions and answers.

# The Remarkable History of Geography and Cartography

Geography is intimately connected with the history and nature of the world, down through the ages from earliest times. Cartography, or mapmaking, which goes hand in hand with the study of geography, antedates even the art of writing. Diagrams of regions that were familiar to them were made by ancient peoples such as the Babylonians, the Egyptians, the Chinese, and, of course, the Greeks, who excelled above other nations in this skill well into the sixteenth century.

These initial questions and answers should be a good way to test yourself to see how much you know about this planet we live on and how places on the map originated and were first recorded.

**❶** The word *geography* means "earth description," a term that evolved during an ancient era when the location of the earth's features and the reasons for natural phenomena were of prime importance to a famous seagoing, much-traveled people. Can you name the country where this word originated?

**❷** The earliest known map is a clay tablet that dates from about 2300 B.C. It shows land features and a river in an area of Mesopotamia (present-day Iraq). What ancient people are credited with making this map?

**❸** The earliest seagoing navigators on record circumnavigated the entire continent of Africa by ship as early as 600 B.C. Who were these seafarers, and for extra credit, where would we locate their country on today's maps?

**❹** One of the first Greeks to interpret geography as a science that could be studied and utilized for the benefit of mankind was Herodotus. He believed the earth was a disk, not a globe, and he explored and wrote about peoples along the coast of Asia Minor. What part of the world does Asia Minor refer to?

**❺** Another well-known Greek was the first person to demonstrate convincingly that the earth was round. Who was he?

**6** This noted Greek based his belief that the earth was round on two natural phenomena he had carefully observed. Can you name them?

**7** The greatest figure in the ancient world in the advancement of geography and cartography lived in the second century A.D. in Alexandria. Who was he?

**8** The famous ancient Greek geographer in the preceding question developed a method of graphic presentation of detailed areas and features on a map or chart. What is this called?

**9** This same Greek geographer also showed how navigators could determine their positions more easily by dividing the globe into segments through the use of vertical and horizontal lines. What are these called?

**10** The ancient Polynesians devised ingenious charts that were constructed of wooden frameworks to which were attached strings that depicted waves and currents and shells that represented islands. Using these charts, they could sail in their dugouts for hundreds of miles. In what part of the world did the Polynesians sail?

**11** It was remarkable how far many of the early peoples could travel in their small ships. The Norsemen, for example, traveled clear across the northern Atlantic Ocean and discovered what is now known as the largest island in the world, lying on the Arctic Circle and covered with snow and ice. What is this land?

**⑫** The Vikings were said to have used compasses with magnetized needles in the twelfth century when they raided the coasts of Europe and the British Isles in their long boats. For this two-parter: (a) where did the Vikings sail from on such raids, and (b) on what sea were their coastal villages located?

**⑬** One of the most famous travelers in history was an Italian from Venice who did much to tell people about the geography of the world as it was during his lifetime in the fourteenth century. He wrote about his lengthy travels in the Far East, describing the terrain, features, and cities he visited. Name him.

**⑭** What was called "The Age of Discovery" was climaxed by the voyage of Columbus in 1492. A popular misconception about this voyage is that he landed on the continental shores of America. Where did Columbus really land?

**⑮** One of the most significant activities during the Age of Discovery was the pioneering of a Portuguese navigator who led a fleet of four vessels around the Cape of Good

Hope and across the Indian Ocean to India. Do you know: (a) who this daring navigator was, and (b) what ocean he sailed to get from Portugal to the Cape of Good Hope?

**16** The Pawnee Indians were said to have impressed the Spanish explorer, Francisco Coronado, when he visited their territory in 1541. They were able to travel at night by using information painted on elk skins. What kind of information was on these native maps?

**17** After the Civil War, John Wesley Powell, an American geologist, explored the Grand Canyon and the famous river that flows through it. Can you name this river that actually created the Grand Canyon?

**18** The American Geographical Society, the first geographical society in the United States, was founded in 1852 in New York. Name the other well-known society in this field, whose magazine is recognized internationally for its superb photos of people and places in the far corners of the world.

**19** This sixteenth-century Portuguese navigator's fleet was the first to sail around the world, completing the voyage in September 1522. Who was he?

**20** The first map showing the Western Hemisphere and decribing a land called "America" was printed in Alsace by a German cartographer, Martin Waldseemuller. About how long ago was the map published: (a) 200 years, (b) 500 years, or (c) 800 years?

**21** In 1588 the first map of an American city appeared in a book about the explorations of Sir Francis Drake in the New World. For this two-parter: (a) what is the name of the city, and (b) where is it located?

**22** Edwin Howell devised the first map of the Grand Canyon that visualized the vertical dimensions of its areas (cliffs, valleys, etc.) What is this type of map called?

**23** In 1728 a Danish seafarer explored the northern Pacific Ocean and sailed between what are now Siberia and Alaska to reach the Arctic Circle. For this two-parter: (a) who was the explorer, and (b) what body of water is named after him?

**24** In 1807 an American trapper named John Colter became the first nonnative to view the wonders of what is now known as Yellowstone National Park. What three states share this park?

**25** Another American pioneer, Thomas Hutchins, discovered Mammoth Cave, so named because of its immense size and now recognized as one of the largest in the world. Where is the cave located?

**26** James Bridger, a fur trader and trapper, was traveling in the West in the early part of the nineteenth century when he came upon an inland body of water that was unique because it was saltier than the ocean itself. What is the name of that body?

**27** During the first decade of the nineteenth century, two very determined Americans spent several years exploring and mapping unknown territory. Although they were considered to have gone "West," much of their mapping occurred in what was later called the Midwest. Can you name these two explorers?

**28** British explorer Hugh Clapperton achieved fame as one of the first Europeans to visit Lake Chad in north central Africa. He later sought to discover and map the mouth of the Niger River. In what countries are these two geographical features located?

**29** No one had ventured far into the interior of Australia until a Scotsman, John McDouall Stuart, traversed the continent in 1862. He provided basic information about the landforms and other features he saw en route. Between what two major oceans does Australia lie?

**30** Despite the huge size of Antarctica, this frigid continent was virtually unknown until two British mariners, William Smith and James Branfield, made rough maps of its northernmost peninsula in 1820. Which of the continents lies closest to Antarctica?

**31** For generations, European and American explorers tried in vain to locate a water route shortcut from Europe to the Far East. It was finally reached in 1851 by an Irish explorer, Robert McClure. What is this water route called?

**32** Sir Henry Morton Stanley uttered the now familiar line, "Dr. Livingstone, I presume," when he found the missing doctor at what African lake, the second largest on the Dark Continent?

**33** Another question on Dr. Livingstone: What two waterways did this Scottish missionary and explorer discover in Africa?

**34** Most people know that when it is winter at the North Pole, it is summer at the South Pole. But do you know: (a) who discovered the North Pole, and (b) who discovered the South Pole?

**35** One of the great Spanish explorers in the New World during the Age of Discovery was Vasco Nuñez de Balboa who navigated the waters around Central America and northern South America. But, in 1513 he completed an historic land crossing between two major bodies of water. For double credit: (a) what was the land he crossed, and (b) what were the two bodies of water?

**36** Another intrepid Spanish explorer was Hernando Cortez, who also visited Central America but was famous for his ventures farther north. With what North American country is his name most closely associated?

**37** Still another Spanish explorer was Ponce de Leon, who was searching for a mythical "fountain of youth." He never found it, but in 1515, after visiting Cuba, he explored a land he erroneously thought was another Caribbean island. What was this land?

**38** Early explorers in South America were astonished at the accomplishments of the people in the Andes who had developed an immense pre-Columbian civilization and built great temples. What were these people called?

**39** In times past, maps produced in Arabian countries were markedly different from Western maps. What was the difference?

**40** One of England's most noted navigators in the late sixteenth century was John Davis, who navigated the waters around Greenland and Labrador, as well as on the opposite side of the globe. He was killed in 1605 fighting Japanese pirates in yet another area of the world, the East Indies. Where are the East Indies?

**41** Sir Walter Raleigh traveled widely during his lifetime as an emissary of British royalty. He is best known to Americans for his explorations during 1585 to 1587 in the region now occupied by one of America's oldest states. Name the state.

**42** In 1803 the United States increased its area by some 828,000 square miles when it made a deal with France to acquire land extending from the Mississippi River to the Rocky Mountains and from Canada to the Gulf of Mexico for the sum of $15 million. What is the name of this historic land deal?

**43** Marquette, Michigan, was named in honor of a famous French missionary, Jacques Marquette, who explored the region in the late seventeenth century. Who was the other noted French explorer who accompanied him?

**44** Another French explorer who was active during this era (for whom a city in Illinois was named) is remembered mainly because of his explorations down the Mississippi in 1682, when he named the region near its mouth Louisiana. Name him.

**45** Henry Hudson was commissioned by the Dutch East India Company in 1609 to find the Northwest Passage. He failed, but he did become the first white man to sail up the Hudson River in a ship whose name became as famous in New York history as that of Hudson himself. What was the name of this ship?

**46** Aerial photography was an important milestone in mapmaking after the invention of the airplane. What invention developed this graphic technology even further?

## Answers

**①** *Greece. One of the earliest descriptions of the world in which the ancient Greeks lived was expressed in the poetry of Homer, author of the* Iliad *and the* Odyssey, *written during the eighth century,* B.C.

**②** *The Babylonians made the earliest map specimen thus far discovered, although certain land drawings found in Egypt and paintings discovered in early tombs are nearly as old. It is quite probable that the Babylonians and the Egyptians developed their mapping skills more or less concurrently.*

**③** *The ancient Phoenicians. Phoenecia was located on the shores of the Mediterranean Sea near what is now Lebanon.*

**④** *This is a great peninsula in western Asia that is washed by the Black Sea* *in the north, the Mediterranean in the south, and the Aegean Sea in the west. It is occupied today by Turkey, sometimes referred to as Anatolia, or the Asian part of Turkey.*

⑤ *Aristotle.*

⑥ *(1) That the shadow of the earth reflected on the moon during an eclipse was always* circular, *and (2) that the horizon always re-*treats *as one walks or rides toward it.*

⑦ *Ptolemy. He systema-tized the study of the earth by compiling an eight-volume* Guide to Geography, *which included every-thing that was known about the earth to that date.*

⑧ *Topography.*

⑨ *Latitude and longitude.*

⑩ *A vast area of the cen-tral Pacific Ocean, extending from New Zealand to Hawaii.*

⑪ *Greenland. It was dis-covered by Eric the Red, the Norse chieftain who was exiled and sailed 1,000 miles to Iceland, and later some 500 more to Greenland to establish a colony where he and his followers could be free.*

⑫ *(a) They sailed from Scandinavia. (b) Their ports were mainly on the North Sea.*

⑬ *Marco Polo.*

⑭ *Columbus landed on what is now San Salvador, in the Bahama Islands, off the southeastern tip of Florida.*

⑮ *(a) Vasco da Gama. (b) From 1497 to 1499 he sailed from the North Atlantic, where Portugal is located, to the South Atlantic where the Cape of Good Hope is situated at the southern tip of Africa.*

⑯ *The positions of the stars overhead.*

⑰ *The Colorado River.*

⑱ *The National Geographic Society, which was founded in 1888.*

⑲ *Ferdinand Magellan. Although his fleet completed the voyage, he never made it. He was killed in a battle in the Philippines in April 1521.*

⑳ *(b) The map was printed almost 500 years ago (1507) when Waldseemuller compiled information, much of it hearsay, from navigators and explorers who claimed to have knowledge of the New World.*

㉑ *(a) St. Augustine, and (b) on the east coast of Florida in the northern part of the state.*

㉒ *A relief map. Five years before rendering the Grand Canyon map, Howell had depicted in relief the island of Santo Domingo in the Dominican Republic.*

㉓ *(a) Vitus Bering, and (b) the Bering Strait.*

㉔ *Though largely in Wyoming, Yellowstone also has fringes in Montana and Idaho.*

㉕ *In southern Kentucky.*

㉖ *Great Salt Lake, Utah. About 11,000 years ago the lake was much larger and deeper, and far less salty. But as the waters receded and evaporated, the salt content from underlying deposits of chemicals and common table salt increased proportionately and continues to do so.*

㉗ *The explorers were Meriwether Lewis and William Clark.*

㉘ *Lake Chad is mainly in Chad, but also stretches into Niger, Nigeria, and Cameroon. The mouth of the Niger River is in Nigeria, at Port Harcourt on the Atlantic Ocean.*

㉙ *The Indian Ocean to the west and the South Pacific Ocean to the east.*

㉚ *South America.*

㉛ *The Northwest Passage. It runs between the Atlantic and the Pacific oceans in northern Canada and along the northern coast of Alaska.*

㉜ *Lake Tanganyika. It is 420 miles long and covers about 12,700 square miles.*

㉝ *Dr. Livingstone discovered both the Zambesi River and Victoria Falls.*

㉞ *(a) Admiral Robert E. Peary discovered the North Pole in 1909, and (b) Roald Amundsen discovered the South Pole in 1911.*

㉟ *(a) The land he crossed was the Isthmus of Panama, and (b) the two bodies of water were the Pacific Ocean and the Caribbean Sea.*

㊱ *Mexico, where he once climbed Mount Popocatepetl, which, though dormant today, at that time was an active volcano.*

㊲ *Florida.*

㊳ *The Incas. They were located largely in Peru, with villages stretching north into Ecuador and south into Chile.*

㊴ *South was always at the top of the map and north at the bottom.*

㊵ *In southeast Asia, but in Davis's time largely referring to India.*

㊶ *Virginia, where he helped establish colonies.*

㊷ *The Louisiana Purchase. Eventually all or parts of fifteen states were formed out of the region.*

④ *Louis Joliet.*

④ *Robert Cavelier La Salle. He named the region in honor of King Louis XIV of France.*

④ *The* Half Moon.

④ *The rocket, which made it possible to place satellites in space from which photographs can be taken of every corner of the globe.*

# Maps and
# Their Meaning

D*espite the avail-ability of atlases of all sizes and kinds, and the proliferation of maps produced by various types of organizations, the average map reader seems to be all thumbs. Many have trouble distinguishing north from south or east from west, not to mention any comprehension of distances and symbols.*

*How much do you think you know about the basics of map reading, let alone cartography? Let's find out.*

❶ What is *cartography*: (a) a comprehensive knowledge of the world and the location of places in it, (b) experience and skill in producing publications relating to the location of points on earth, or (c) the graphic and conventionalized representation of spacial phenomena on a plane surface?

**2** When looking at a map to determine distances, you also may want to know something about the topography of the land, that is, the heights and depths of mountains and valleys and other features. What kind of map has such information?

**3** What system is used on practical reference maps (such as road maps) to help you find a place or location when you look it up in the index?

**4** Is the top of the map north or south?

**5** This instrument, commonly installed in automobiles to measure the distance traveled, is valuable in determining data relating to maps. What is it called?

**6** In the second century A.D., Ptolemy divided the world into imaginary lines running from north to south and east to west. Is it latitude that runs north to south, or longitude?

**7** Why is Greenwich, England, important to maps of the world, or any map, in fact, that makes use of lines of latitude and longitude?

**8** When we speak of "parallels" on a world map what are we talking about?

**9** Lines on a map that are similar to lines of longitude are called: (a) lines of direction, (b) coordinates, or (c) meridians?

**10** Long before computer icons came into use, maps had graphic symbols that were familiar to map readers. Interpret the symbols from a city map (at the left) with their meanings (at the right):

| Symbol | Meaning |
| --- | --- |
| **1.** &#9855; | **(a)** Airport |
| **2.** 🌲 | **(b)** Museum |
| **3.** ✈ | **(c)** Library |
| **4.** 📖 | **(d)** Hospital |
| **5.** ⌶ | **(e)** Park |

**11** Topographical maps often use colors to help readers identify the major features shown. Match the features on the left with the colors most commonly used, on the right:

| Feature | Color |
| --- | --- |
| **1.** Man-made features | **(a)** Blue |
| **2.** Elevations | **(b)** Green |
| **3.** Water | **(c)** Black |
| **4.** Vegetation | **(d)** Brown |

**12** Magnetic compasses were invented by unknown navigators some time in the tenth century. They made use of the fact that a metal needle would be influenced by magnetic polar forces so that it always pointed north. How many degrees are there in a compass?

**⓭** On a standard road map: (a) what is the name of the key that explains the symbols being used and (b) what are the numbers and alphabetical letters called that pinpoint a given position?

**⓮** We have all seen decorative circles printed on maps to show points of the compass. What are these decorative circles called?

**⓯** If you see a map of the world that looks like the skin of a grapefruit that has been pressed flat, what is this strange graphic representation?

**⓰** Maps often exaggerate the size of certain features in order that they may be more readily seen and understood by viewers. What is the name for the device printed on such maps that indicates how much exaggeration has been used?

**⓱** Maps showing the whole world or large portions of it usually have a gauge in the lower right-hand corner that shows that the scale at the equator may be much different from the scale at, say, 70° N latitude. What is the name for this gauge?

**⓲** Even ordinary road maps use certain kinds of exaggeration to make themselves clear. Cities and towns are usually more spread out than they would be if precisely shown, while major highways and parkways are wider. How can you be sure, therefore, that the distances are accurate when you are planning a trip in the family car?

**⑲** A terrestrial map is one that shows the earth. What is the name for a map that shows the heavens?

**⑳** If you want to obtain accurate mileages on a map without having to add a list of distances, you can use an instrument with a mileage gauge and a small wheel that you roll along a map. What is this instrument called?

**㉑** The work of surveyors in the preparation of detailed maps is critical. What is a surveyor?

**㉒** An agency of the United States government is called the U.S. Board of Geographic Names. It is responsible for identifying the correct name of every geographic feature (lake, city, river, island, mountain, etc.) in the world, which mapmakers and governments use in their maps and publications. How many place names do you think it has on record: (a) 300,000, (b) 2,500,000, (c) 3,300,000, or (d) more than 4,000,000?

**㉓** Other specialists who are important to the study of geography are demographers. A demographer is a person who: (a) is an expert in the use of colors and tones for maps, (b) compiles information about the world's climates, or (c) gathers data and compiles vital statistics on people and populations?

**24** Maps can be as detailed or specialized as the cartographer desires. Listed below are some common types of specialized maps. Try to match them with the brief descriptions:

| Map | Description |
|-----|-------------|
| **1.** Flow map | **(a)** Shows capitals and borders of counties, urban areas, and other sectors |
| **2.** Political map | **(b)** Indicates subdivisions and property lines, particularly for real estate use |
| **3.** Geologic map | **(c)** Contains traffic routes and data for various forms of transportation |
| **4.** Topographic map | **(d)** Reveals structures of the earth as though the soil were stripped away |
| **5.** Cadastral map | **(e)** Shows relative heights and depths of land features in regions covered |

**25** How well can you read a simple map? To find out, answer these five questions by referring to the sample map:

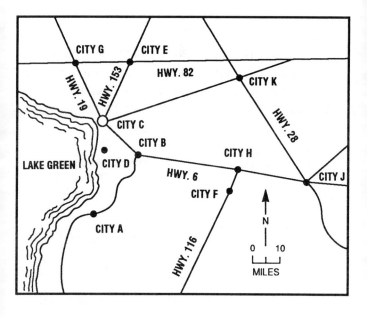

(a) What is the westernmost city on this map?

(b) Highway 82 passes through which two cities on the map?

(c) It's a little over one inch from City J to City K on this map. About how many miles actually separate the two cities?

26

(d) To travel from City B to City G, you would need to go in what direction?

(e) On this map, which city do you think has the best harbor and port facilities?

**26** Symbols were devised to make it easy for readers to get a quick, visual impression of topographical and man-made features depicted on maps. Try to match the symbols on the left with their meanings:

| Symbol | Meaning |
|--------|---------|
| **1.** Circle with a star | **(a)** Federal highway |
| **2.** Square with a flag | **(b)** Church |
| **3.** Square with a cross | **(c)** Railroad tracks |
| **4.** Line with crossbars | **(d)** School |
| **5.** Shield with a number | **(e)** Capital city |

**27** The panel on a standard road map that shows the towns and cities included is called: (a) an index, (b) a legend, or (c) a scale.

**28** Most current maps designate distances on their scales in kilometers as well as miles. For this two parter: (a) which is longer, a mile or a kilometer, and (b) precisely, how do they compare in length?

**29** People who use maps for hiking refer to their use as *orienteering*. Does this term mean: (a) judging walking distances from the east, or "orient" portion of the map, (b) developing one's skill at map reading to make the most effective use of the information at hand, or (c) making notes on the map about the topography while en route?

**30** If you were trying to find your way through a large city and were confused about directions, what would the following have in common that might help you: one side of the buildings has more moss and ivy than the other; there is more snow on one side of the buildings than the other; the paint is less faded and peeled on one side of the buildings than the other?

**31** Many generations ago, navigators learned how to make sure their charts were placed in the right position by observing the stars at night. Of great importance was the constellation known as Ursa Major, which contains two stars that point to the North Star. What is the popular name for this constellation?

**32** If you are using a map and a compass at the same time, you will notice that the arrow marked "North" on the map is not quite in line with the northerly direction pointed by the compass needle. This variation is caused by the difference between true north (the map direction) and magnetic north (the compass direction). In geographic terms, what is this variation called: (a) declination, (b) orientation, or (c) protraction?

**33** What are the "cardinal points" of the compass?

**34** If you wanted a map with "hydrographic" features clearly shown, what would you be hoping to see on it?

**35** Can a printed map have more than one scale?

**36** What agency in the United States is responsible for preparing topographic and geological maps?

**37** What government department is responsible for the agency in the preceding question?

**38** Most globes of the world that you see on stands or sitting on tables and desks are absolutely spherical in shape. Why is this misleading?

**39** A map that shows the depths of the sea in any given area is known as: (a) an isometric projection, (b) a bathymetric chart, or (c) an acquiferous map?

**40** Having progressed this far in *Whiz Quiz/Geography*, you should now be able to tell how well you are acquainted with the world in which you live. Here's a chance to find out by answering a map question from a recent test sponsored by the National Geographic Society. The test was given to nearly 12,500 adults in ten nations, including the United States, Sweden, West Germany, Canada, Japan, Italy, France, the United Kingdom, the Soviet Union, and Mexico.

*Where In The World Are You?*
Match the numbers on the world map on page 31\*
with these places:

| | |
|---|---|
| _____ United States | _____ United Kingdom |
| _____ Soviet Union | _____ South Africa |
| _____ Japan | _____ West Germany |
| _____ Canada | _____ Egypt |
| _____ France | _____ Vietnam |
| _____ Mexico | _____ Central America |
| _____ Italy | _____ Persian Gulf |
| _____ Sweden | _____ Pacific Ocean |

---

\*The world map and this question are reprinted by permission of the National Geographic Society and the Gallup Organization.

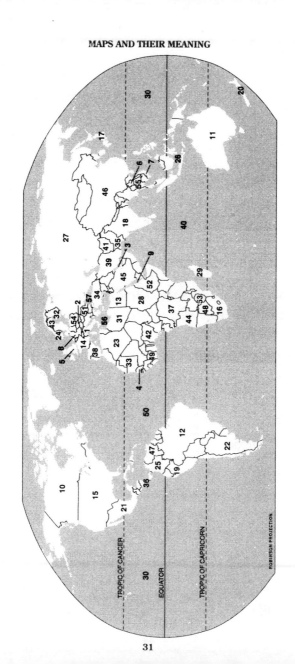

ROBINSON PROJECTION

TROPIC OF CANCER

EQUATOR

TROPIC OF CAPRICORN

# Answers

① *(c) The representation can be as selective or as broad as the cartographer desires to show quantitative and qualitative data.*

② *A contour map, also called a relief map.*

③ *The map is divided into squares, referred to as "grids," created by uniformly spaced vertical and horizontal lines that form coordinates where they meet.*

④ *North is at the top of the map ninety-nine times out of one-hundred, shown by an arrow that may or may not be marked "N." When there is any reason for the map to be oriented otherwise, the direction will be clearly marked.*

⑤ *An odometer.*

⑥ *Longitude lines run north to south; latitude lines run east to west. Latitude is the relative "northness" or "southness" of points on the earth's surface, measured in degrees of arc from the equator, from zero degrees at the equator to ninety degrees at the poles. Longitude is the relative "eastness" or "westness" of points, measured in degrees of arc from a north-south line.*

⑦ *Greenwich, England, a borough of London, is the former site of the Royal Observatory and was selected by an international convention as the location of the prime meridian which is zero degrees longitude, from which all other longitudes are measured.*

⑧ *Parallels are lines that run parallel to the equator and are the same as lines of latitude.*

⑨ *(c) Meridians.*

⑩ *1(d), 2(e), 3(a), 4(c), and 5(b)*

⑪ *1(c) Man-made/Black, 2(d) Elevations/Brown, 3(a) Water/Blue, and 4(b) Vegetation/Green.*

⑫ *360 degrees.*

⑬ *(a) The symbols key is called a* legend, *and (b) the numbers and letters that pinpoint a position are called* coordinates.

⑭ *They are called compass roses. Some are quite ornate.*

⑮ *This is what is called an "interrupted projection." It is indeed what would result if you were to take a globe of the world, split one side of it in half from pole to pole and then squash it flat. The distortion is reduced (by comparison with a flat map) by dividing the world into segments.*

⑯ *The vertical scale. Like the horizontal scale used to indicate the number of miles per inch, it lets the reader know the true heights or depths of the features.*

⑰ *Mercator projection.*

⑱ *The distances between towns are marked with small numbers.*

⑲ *An astronomical chart, also called a celestial chart.*

⑳ *A cyclometer. The gauge can easily be set to match the scale on the map being used.*

㉑ *A surveyor is a person who uses instruments and mathematics to determine the exact positions and boundaries of tracts of land, roads, and other topographical features.*

㉒ *More than 4,000,000.*

㉓ *(c) A demographer keeps tabs on population data and changes.*

㉔ *1(c), 2(a), 3(d), 4(e), and 5(b).*

㉕ *(a) City G, (b) Cities E and G, (c) 40 to 45 miles, (d) northwest City C.*

㉖ *1(e), 2(d), 3(b), 4(c), and 5(a).*

㉗ *(a) An index.*

㉘ *(a) A mile. (b) A kilometer is .6 mile; conversely, a mile is 1.6 kilometers.*

㉙ *(b) In recent years, orienteering has become quite competitive, the basis for hiking contests.*

㉚ *These features tend to be found on the* north *side of a building, thus helping you to determine your compass direction.*

㉛ *The Big Dipper.*

㉜ *(a) Declination.*

㉝ *The four cardinal points are North, East, South, and West.*

㉞ *You would expect to see water features such as rivers, streams, lakes, and ponds. The word is derived from the Greek* hydro *("water") and* graphein *("to write").*

㉟ *Only if one or more inserts or panels are used, each with its own scale.*

㊱ *The United States Geological Survey. In 1990 alone, the USGS completed a series of 57,000 maps based on aerial photographs.*

㊲ *The Department of the Interior is responsible for the USGS.*

㊳ *The world is not completely spherical, but somewhat flattened at the poles.*

㊴ *(b) A bathymetric chart shows sea depths.*

㊵ *Americans could identify an average of only 8.6 of the 16 places shown on the blank world map. The Swedes were at the top of the ten nations polled, with an average of 11.6 correct. The answers:*

## MAPS AND THEIR MEANING

# A World Tour

Tours have long been popular for Americans of all ages. Thanks to the promotional efforts of transportation companies, travel agents, and thousands of Chambers of Commerce, we are visiting historical sites, enjoying natural wonders, and exploring small hamlets and large cities alike.

Only one thing does not seem to keep pace with all of this movement around the face of the globe: the average person's knowledge of geography. Where in the world are they? Many don't seem to know at all. Now that we have sampled some of the history relating to geography, let's see how well you could locate yourself on an imaginary tour of our planet.

**①** The Isle of Capri is noted for its dramatic natural wonders, such as the Blue Grotto, wind-sculptured peaks, and arches of rocks. What tourists admire as scenic wonders, geologists study as examples of how the forces of nature create and shape the world we live in. Where would you go if you wanted to take a tour of Capri?

**②** You have always loved ancient history and have joined a tour that has placed you in the perfect position for a photo atop the largest pyramid in the world. Where are you?

**③** How would it be possible for you to travel less than five miles from a border of the United States and be in a foreign country other than Canada or Mexico?

**④** If you were to fly in a straight line directly south from Boston to the tip of South America, what countries would you pass over in flight?

**⑤** Most tourists who go south are familiar with the equator, the imaginary great circle around the earth. But where would you be if you were at a latitude of almost twenty-four degrees north of the equator, which is the farthest point north at which the sun can be seen directly overhead at noon?

**6** Where would you be if you were at a similar parallel almost twenty-four degrees *south* of the equator?

**7** You have gone skiing in Switzerland and in the course of your travels there have seen the Simplon, the Brenner, and the Saint Bernard. What are these places?

**8** The northernmost point on the east coast of Australia is barely 100 miles from a tropical island nation that lies to the north and almost on the equator. If you were to cross over to this island by boat, where would you be?

**9** You do not have to leave the earth to visit the Mountains of the Moon. But where would you have to fly to in order to see them?

**10** You find yourself in Nome, Alaska, which is located on the Seward Peninsula at the westernmost tip of the state. Flying from Nome westward across the Bering Strait, you would be over what foreign country in about fifteen minutes?

**11** The Great Barrier Reef is more than 1,200 miles long and is said to contain almost 500 types of coral and 1,400 species of fish. If you were an ardent scuba diver and wanted to experience diving along this, the largest coral reef in the world, what country would you fly to? For extra points, what state or territory is the reef closest to?

**⑫** If you were to fly due west from Dublin to North America you would reach the lower region of Hudson Bay in Canada, which is at the same latitude as Ireland. Why then is Ireland's climate so mild?

**⑬** Some years ago, a humorous travel book was written entitled *Lost in the Horse Latitudes*. What are these latitudes?

**⑭** These place names all have one thing in common: Kalahari, Erg Chech, Takla Makan, Gibson, and Atacama. What are they?

**⑮** You tell a friend that you are going to Paris and will visit the Eiffel Tower, which you think is the tallest in the world. But you find out there is a much taller tower in Canada, where you can dine in a restaurant at a height of 1,140 feet. What is this stucture?

**⑯** One country you have decided to visit while in Europe has been described as having "the highest population density on the continent." But you won't mind the crowds because the country is tiny, very cosmopolitan, and known for attracting the rich and famous. What is its name?

**17** If you wanted to visit a city within the Arctic Circle with a population of several hundred thousand people to see how the inhabitants survive in extreme cold, you would not have to struggle to make a choice because there is only one such metropolis in the world. What is its name and where is it?

**18** You are leaving on a trip to the Pyrenees Mountains in northern Spain and southwestern France. What is the name of the people who live there?

**19** You are going to visit India and surrounding lands and have been told that one of the most dramatic moments of the trip will be driving through the Khyber Pass. Where is this pass?

**20** If you found yourself in the city of Agra in Uttar Pradesh state in northern India, what world-famous mausoleum would you have come to see?

**21** You are traveling in Asia and have decided to visit the Forbidden City. For a two-parter: (a) where is the Forbidden City, and (b) why was it so named?

**22** You would like to visit at least a little part of Russia but don't like cold climates. Your travel agent assures you that you can see the southern region of the Soviet Union and be no farther north than if you were in the mid-eastern area of the United States. What much-visited city on the east coast of the United States has about the same latitude as Russia's southern border?

**23** How many countries can you visit without ever going outside the United Kingdom (not counting possessions), and what are they?

**24** A tour you are considering indicates that it would take you through the Low Countries. What are the Low Countries, and why are they so named?

**25** If you were lunching on the banks of the Thames River watching international rowing and sculling races, in what small English town would you be?

**26** You are flying to Europe and are determined that one place you *won't* go is to the world's busiest international airport. What airport would you avoid?

**27** You have been invited by a friend north of the border to go on a fishing trip to the smallest province in Canada, provided you can identify it accurately. What would your answer be?

**28** This is the life! You are sailing in a small boat made of reeds across the world's highest large lake, at an altitude of

12,500 feet, with one of the most magnificent mountain views you could ever imagine. What is the lake?

**29** Like most visitors to China, you go for a stroll along the Great Wall and are told that it stretches for some 1,500 miles. How long did it take to build this serpentine wonder?

**30** If you were to visit its capital, San Jose, you would hear the people referred to as *ticos*. What Central American country would you be in?

**31** You want to visit the Caribbean islands that were bought by the United States from Denmark and are now popular tourist centers. Where will you go?

**32** A flight cannot be booked to Gondwana Land because no such place actually exists in the world. Why then is it important in the study of geography and cartography?

**33** If you want to go to Canada but you don't like cold weather, you would select the southernmost city to visit. What is this city?

**34** You are planning to visit a popular Canadian tourist region in the west that is not only beautiful but noted for its mild climate and lush green foliage, even though it lies farther to the north than any of the states in the United States. Where are you going?

**35** You have signed up to join a unique expedition to try to track down and take photos of the Abominable Snowman, described as a shaggy, two-legged creature about seven feet tall and covered with hair. To what part of the world will you be going?

**36** While on a trip to Europe you would like very much to see Stonehenge, with its circles of ancient stone pillars. You are told it is located on Salisbury Plain. Where is that?

**37** If you were to visit (a) the largest, and (b) the smallest countries in South America, where would you go?

**38** With the completion of the tunnel under the English Channel, will tourists be able to drive their own cars between England and France?

**39** Match these world-famous art galleries with the cities in which they are located:

| Museum | City |
|---|---|
| **1.** The Prado | **(a)** Leningrad |
| **2.** The Hermitage | **(b)** Madrid |
| **3.** The Tate Gallery | **(c)** Lisbon |
| **4.** Gulbenkian Foundation | **(d)** Paris |
| **5.** Musée d'Orsay | **(e)** London |

**40** Your trip itinerary calls for you to spend two days in the "Venice of the North." In what European city will you be staying?

**41** What country would you have to visit if you wanted to stay entirely within its boundaries yet see the most frontiers of any country in the world—thirteen in all?

**42** Try another matching question. Match these features that attract tourists and other travelers with the cities in which they are located:

| **Attraction** | **City** |
|---|---|
| **1.** The Wailing Wall | **(a)** Venice |
| **2.** The Casbah | **(b)** Quebec |
| **3.** Alcazar Castle | **(c)** Jerusalem |
| **4.** Bridge of Sighs | **(d)** Segovia |
| **5.** Shrine of Saint Anne de Beaupré | **(e)** Algiers |

**43** If you wanted to see the city that is known as the northernmost capital in the world, where would you go?

**44** Punta del Este is a very popular seaside resort town on the southern coast of Uruguay in South America. It has a normal population of about 15,000, but in January this figure swells to as high as 350,000. Why in January?

**㊺** If you wanted to see the Alhambra, called "the greatest monument of Islamic Spain," what city would you include in your itinerary?

## Answers

① *To the southwestern region of Italy, below Naples, where you would take a boat to Capri, in the Tyrrhenian Sea.*

② *You are not in Egypt, but on Quetzalcoatl in Mexico, which is almost a third again as large as the great pyramid of Cheops along the Nile.*

③ *If you were to go to the Bering Strait off the Pacific Coast of the United States you could go only three miles from an American-owned island to one that belongs to Russia.*

④ *In order, from north to south, you would fly over the Dominican Republic, Venezuela, Colombia, Peru, Brazil, Chile, and Argentina.*

⑤ *The Tropic of Cancer, which marks the northern boundary of the tropics.*

⑥ *The Tropic of Capricorn, which marks the southern boundary of the tropics.*

⑦ *They are mountain passes through the Alps.*

⑧ *In Papua New Guinea, a country slightly larger than the state of California.*

⑨ *You would have to go to east central Africa to the border of Uganda and Zaire. These mountains, which rise to elevations of almost 17,000 feet, are officially called the Ruwenzori.*

⑩ *The USSR is less than 100 miles from the west coast of Alaska at the point where the two continents are the closest together.*

⑪ *You would fly to Australia. The reef lies along the state of Queensland, in the northeastern sector of the country.*

⑫ *The climate is mild in Ireland because it is warmed by the North Atlantic Drift, which is a continuation of the Gulf Stream and affects European climates as far as the north coast of western Russia.*

⑬ *The horse latitudes refer to two belts in the oceans, at about thirty degrees latitude in each hemisphere, where the weather is hot and dry and the water unusually calm. The most familiar location is just south of Bermuda. The name derives from the days when Spanish ships were transporting horses to America and became so becalmed and low on fresh water that the crew had to throw some of the horses overboard to survive.*

⑭ *They are deserts, and you probably would not want to be in any of them since they are among the most desolate and hostile in the world.*

⑮ *The CN Tower in Metro Centre, Toronto, which rises to a height of 1,822 feet.*

⑯ *Monaco.*

⑰ *Murmansk, in the northwestern tip of Russia on the Barents Sea. A major port with a harbor that is astonishingly ice free, it is home to some 350,000 people.*

⑱ *The Basques, the oldest surviving ethnic group in Europe. The people speak a distinctive language that has eight dialects and is not directly related to any other existing language.*

⑲ *The pass, which reaches an altitude of 3,500 feet, crosses from Peshawar, Pakistan, to Kabul, Afghanistan. Besides a modern road, the pass has a railroad that runs through thirty-four tunnels and over ninety-two bridges.*

⑳ *The Taj Mahal, considered one of the most beautiful buildings in the world.*

㉑ *The Forbidden City is (a) located in Beijing (Peking), capital of China, and (b) was so named because it was the residence of the emperor and therefore off-limits to all but a select few.*

(22) *Russia's southern border is at about the same latitude as Washington, D.C.*

(23) *Four in all: England, Scotland, Wales, and Northern Ireland. The Republic of Ireland (the southern area) is not part of the United Kingdom.*

(24) *The Low Countries include the Netherlands, Belgium, and Luxembourg. They were so named because they are flat and near sea level.*

(25) *You would be in Henley, site of the Royal Henley Regatta, held each spring and attracting crews from all over the world to row in singles, doubles, fours, and eights.*

(26) *The airport is Heathrow, outside of London.*

(27) *Prince Edward Island, which is only slightly larger than the state of Delaware. It lies off the east coast, just north of Nova Scotia.*

(28) *Lake Titicaca, which straddles the border between Bolivia and Peru and stretches for 110 miles.*

(29) *The first unified wall was started in the third century, B.C., to protect China from northern nomads. But construction was intermittent until the Ming Dynasty when laborers*

were conscripted from all over China to build it. Even with this accelerated program, it took almost 300 years to complete the Great Wall as we know it today.

③⓪ *Costa Rica.* The term tico *refers in a complimentary way to the size of the citizens, who are generally slight in stature.*

③① *The Virgin Islands, which were purchased in 1917 for $25 million mainly because of their strategic position along the route to the Panama Canal. At that time, the American government had no interest in developing a tropical retreat.*

③② *This was the name given to a huge land mass that supposedly existed in prehistoric times before the continents "drifted" apart to form the ones as we know them today.*

③③ *Windsor, Ontario, a city of 120,000 people, which is comparable in latitude to Albany or Boston.*

③④ *Vancouver Island in southwest British Columbia. It has a mild climate because it lies near warm Pacific Ocean currents that minimize winter cold and inhibit snowfalls.*

③⑤ *To the Himalayan Mountains, where the creature has supposedly been spotted for generations, yet never photographed.*

③⑥ *In Wiltshire in southern England.*

③⑦ *(a) Brazil, the fifth largest country in the world, and (b) French Guiana, which is about the size of Louisiana.*

③⑧ *Travelers will be able to ride in their own cars, but not drive them. Cars will be driven onto specially equipped railway trains and transported between terminals located near each end of the tunnel.*

③⑨ *1(b) Prado/Madrid, 2(a) Hermitage/Leningrad, 3(e) Tate Gallery/ London, 4(c) Gulbenkian Foundation/Lisbon, and 5(d) Musée d'Orsay/Paris.*

④⓪ *Stockholm, Sweden.*

④① *China, whose frontiers include Afghanistan, Bhutan, Burma, Hong Kong, India, Laos, Macau, Mongolia, Nepal, North Korea, Pakistan, USSR, and Vietnam.*

④② *1(c) Wailing Wall/ Jerusalem, 2(e) Casbah/Algiers, 3(d) Alcazar/ Segovia, 4(a) Bridge of Sighs/Venice, and 5(b) Shrine of Saint Anne de Beaupré/Quebec.*

④③ *Reykjavik, Iceland, which lies at 64° 08′ N.*

④④ *Because Uruguay lies well below the equator, the seasons are reversed from those in the north and January is midsummer.*

④⑤ *Granada, in the south/ central region of Spain, about thirty-five miles north of the Mediterranean Sea.*

# Place Names on the Map

**G**eography is a compilation of names, names, and more names. Some are descriptive, but many stem from curious and often quite unexpected origins, attesting to the ingenuity, imagination, or whimsicality of the original name-dropper. Here are some questions to test your skills with words and meanings.

**❶** Let's start with "America" itself. How did it get its name?

**❷** The scientific study of place names did not begin until the early part of the nineteenth century, when it was realized that the words often provided clues to lost languages and peoples. What is the name for this scientific study?

**❸** This city, the capital of one of America's thirteen original colonies, owes its name to an error. It was initially named after St. Bartolph's Town near the eastcentral coast of England, but the name was slurred by the British. What is this city called today?

**④** The name of this Michigan city, which begins with a "K" and rhymes with "you," intrigues us because of its humorous spelling and sound. What is it?

**⑤** A well-known city in Montana derived its name from the tableland nature of the terrain near which it was founded. What is this city?

**⑥** Many places in the western part of the United States derived their names from Indian words. This territory, which became the forty-sixth state in 1907, takes its name from the Indian words for "people" and "red." What is the state?

**⑦** This territory, purchased by the United States from the Russians in 1867, was a matter of great controversy because it was considered a wasteland. The Indian interpretation for its name proved to be meaningful: "the Great Land." What is its name today?

**⑧** The Amazon River was given one of the earliest names in the New World when, in 1541, a Spanish explorer, Francisco de Orellana, navigated the river from its origins in the Andes to the Atlantic Ocean, some 4,000 miles in all. He called the river *Rio de las Amazonas* when he was attacked by natives along the banks. What was it about these natives that made him coin that name?

**9** In 1733, when James Oglethorpe founded a settlement on the shores of a wide, navigable river in the southern United States, he named it after the anglicized version of a Spanish word, *sabana*, meaning a treeless plain. What is this city?

**10** This area north of New England was named "New Scotland" by British colonists. By what name do we know it today?

**11** This country was named for one of the largest rivers in Asia, which has its origins in Tibet and empties into the Arabian Gulf. For a two-parter: (a) what is the country, and (b) what river is it named for?

**12** If you were to translate "Gobi Desert" into the language native to the region in which it is located, the people would not know what desert you were talking about. Why is this so?

**13** Sri Lanka is an island southeast of India, only a little more than 25,000 square miles in size and with a population of about 18,000,000. What was the former name of this tiny, mountainous republic? (Hint: think about tea.)

**14** A number of places have been named because of the dates on which they were discovered. One is a tiny island off the coast of Chile that was first sighted on a Sunday of great importance to Christians. What is its name?

**15** What do the Japanese call Japan?

**16** Match these foreign cities with their popular names:

| City | Popular Name |
|------|--------------|
| 1. Aberdeen, Scotland | (a) City of Peace |
| 2. Baghdad, Iraq | (b) City of the Golden Horn |
| 3. Paris, France | (c) Granite City |
| 4. Istanbul, Turkey | (d) Pearl of the Desert |
| 5. Damascus, Syria | (e) City of the Lily |

**17** Now match these American cities with their popular names:

| City | Popular Name |
|------|--------------|
| 1. Nashville, Tennessee | (a) City of Churches |
| 2. Detroit, Michigan | (b) Monument City |
| 3. Brooklyn, New York | (c) City of Rocks |
| 4. New Orleans, Louisiana | (d) City of the Straits |
| 5. Baltimore, Maryland | (e) Crescent City |

**18** It has been called perhaps more names than any of the early cities in the United States: the Athens of America because of the preponderance of Greek architecture in its public buildings, the City of Notions because of the number of notions manufactured there in Yankee trading times, the Hub of the Universe, and the Tri-Mountain City from the three hills on which it was originally built. What American city is this?

**⓳** In many lands, particularly throughout Asia and the Pacific, place names tend to be romantic and poetic. Match the following names and cities:

| City | Poetic Name |
|------|-------------|
| **1.** Hong Kong | **(a)** The Abode of Snows |
| **2.** The Himalayas | **(b)** Land of Lakes |
| **3.** Mindanao (Philippines) | **(c)** Fire God Mountain |
| **4.** Japan's Fujiyama | **(d)** Sweet, Fragrant Waterway |

**⓴** The major city in Hawaii was so named because the word is a derivation of an ancient islander term meaning "Fair Harbor." Which city is this?

**㉑** When the New World was first being explored, the names given to places newly discovered were largely of Indian, Spanish, or Portuguese origin. Not until 1584 was the first *English* name given to a place in America, to identify the colony that would become known as Virginia. From what famous person did this name come?

**22** Match the names of the following places with the descriptive native translation of their names:

| City | Translation |
|------|-------------|
| **1.** Beirut | **(a)** House in the Swamp |
| **2.** Liverpool, England | **(b)** New City |
| **3.** Brussels | **(c)** Noisy Place |
| **4.** Lima, Peru | **(d)** Pool of Thick Water |
| **5.** Naples | **(e)** Wells |

**23** Many place names were coined and added to the map to evoke feelings of optimism, joy, and hope. Among these are the following. Match the names on the left with the states:

| Town | State |
|------|-------|
| **1.** Happy | **(a)** Pennsylvania |
| **2.** Good Hope | **(b)** Utah |
| **3.** Joy | **(c)** Texas |
| **4.** New Hope | **(d)** Louisiana |
| **5.** Merry Hill | **(e)** Arkansas |
| **6.** Paradise | **(f)** Oregon |
| **7.** Pep | **(g)** North Carolina |
| **8.** Pleasant Grove | **(h)** California |
| **9.** Sunny Valley | **(i)** West Virginia |
| **10.** Friendly | **(j)** New Mexico |

**24** Some places are named by mistake. A classic case is that of a certain northwestern American city. On an early map the location of this city was indicated but without any identity. A puzzled cartographer making corrections on the map, pencilled in "NAME?" to try to elicit that detail from his colleagues. This was misread and the name took root. What is the name of this city?

**25** Why is Ireland called "The Emerald Isle"?

**26** When the Spanish explorer, Cordoba, discovered this peninsula, now a part of Mexico, he gave it a name through a misunderstanding. When he asked a native the name of this land, the reply was "I don't understand you" in the native tongue. What is the name of this peninsula, surrounded by the Gulf of Mexico on two sides and the Caribbean Sea on the third?

**27** The name of the capital of Iowa is French for "of the monks." It was erroneously thought to be the habitat of a group of monks because it was called *Moingouena* by a local tribe. Name this city.

**28** Houston, Texas, is just one of many places in America that originated with the name of a person, in this case Sam Houston, a tough frontier hero in the region in the early nineteenth century. What was the nickname given

Texas when it was admitted to the Union during Sam's era in 1845?

**❷❾** Transplanting place names from abroad to America has led to much confusion in the past. You don't have to go outside the United States, for example, to visit some well-known "foreign" cities. In what states would you find the following? Name at least one for each city.

1. Paris          _____
2. London       _____
3. Dublin        _____
4. Madrid        _____
5. Moscow      _____
6. Lisbon         _____
7. Bombay      _____
8. Amsterdam   _____
9. Rome           _____
10. Copenhagen   _____

**❸❿** Even the names of states have been transplanted. Name at least one state in which you would find these "city-states."

1. California    _____
2. Indiana       _____
3. Texas         _____
4. Florida        _____
5. Maine         _____
6. Ohio           _____
7. Oregon       _____

**31** Only about twice the size of Rhode Island and named for an Arabic phrase meaning "Little Fort," this tiny country lies on the Persian Gulf, with Iraq to the northwest and Saudi Arabia to the south. Can you name it?

**Many places on the globe have a *color* in their names. Geographically speaking, how color-conscious are you? Try to answer the following questions to find out.**

**32** This body of water has shorelines in Egypt, Saudi Arabia, Ethiopia, Sudan, and Yemen. Name it.

**33** This body of water touches Turkey, Bulgaria, Romania, and the USSR. Name it.

**34** This body of water is an arm of the Pacific Ocean lying between China and Korea. Its oriental name is *Hwang Hei*. Name it.

**35** One of the major rivers of the world is the Hwang Ho, which flows some 3,000 miles through China. What is its popular Western name?

**36** There are almost forty major land and water features on the globe whose names include the word "blue." What is the largest geographical feature in Virginia that includes this color in its name?

**37** Quite a few geographical features are so colorful that one hue is not enough for their names. Such is the case with the largest natural stone arch in the world, located in southern Utah. What is it?

**38** This South Carolina city is only one of many places in North America named for a King of England. For a two-parter: (a) name the king, and (b) name the city.

**39** Among the Indian names used to designate early settlements in the New World was *Quenticutt*. Name the New England state that is derived from this word, which means "at the river mouth."

**40** The Spanish explorers delighted in creating grandiose names for places they discovered and settlements they established. A good example was a town founded in 1598 by Don Carlos on the Pacific coast of what later became California. He called it *La Villa Real de Santa Fe de San Francisco*, later shortened to Santa Fe. There are a number of Santa Fe names in America. What do the words mean?

**41** There is a famous city in Italy known as "City of the Bull," which at one time was the capital of the Kingdom of Italy. Do you know its name?

# *Answers*

(1) *America was named in error after Italian navigator Amerigo Vespucci, who claimed to have reached what is now the United States mainland even before the arrival of Christopher Columbus.*

(2) *Toponymy.*

(3) *Boston.*

(4) *Kalamazoo.*

(5) *Butte, named after a butte, which is a flat-topped hill, steep-sided and generally rising from a grassy plain.*

(6) *Oklahoma. ("People" is* okla; *"red" is* humma.*)*

(7) *Alaska, which entered the Union as the forty-ninth state in 1959.*

(8) *De Orellana noted that there were a number of women ("Amazons") in the ranks of the attacking natives.*

(9) *Savannah, Georgia.*

(10) *Nova Scotia.*

(11) *(a) India, (b) the Indus.*

(12) *Because* Gobi *means desert and the translation would simply be "desert desert."*

⑬ *Ceylon, exporter of tea, rubber, and coconut oils.*

⑭ *Easter Island.*

⑮ Nippon, *which means "origin of the sun."*

⑯ *1(c) Aberdeen/Granite City, 2(a) Baghdad/City of Peace, 3(e) Paris/City of the Lily, 4(b) Istanbul/City of the Golden Horn, and 5(d) Damascus/Pearl of the Desert.*

⑰ *1(c) Nashville/City of Rocks, 2(d) Detroit/City of the Straits, 3(a) Brooklyn/City of Churches, 4(e) New Orleans/Crescent City, and 5(b) Baltimore/Monument City.*

⑱ *Boston.*

⑲ *1(d) Hong Kong/ Sweet, Fragrant Waterway, 2(a) Himalayas/The Abode of Snows, 3(b) Mindanao/Land of Lakes, 4(c) Fujiyama/Fire God Mountain.*

⑳ *Honolulu.*

㉑ *Elizabeth the First of England, popularly known as "the Virgin Queen."*

㉒ *1(e) Beirut/Wells, 2(d) Liverpool/Pool of Thick Water, 3(a) Brussels/House in the Swamp, 4(c) Lima/Noisy Place, and 5(b) Naples/New City.*

㉓ *1(c) Happy/Texas, 2(d) Good Hope/Louisiana, 3(e) Joy/Arkansas,*

*4(a) New Hope/Pennsylvania, 5(g) Merry Hill/ North Carolina, 6(h) Paradise/California, 7(j) Pep/New Mexico, 8(b) Pleasant Grove/Utah, 9(f) Sunny Valley/Oregon, and 10(i) Friendly/West Virginia.*

㉔ *Nome (Alaska).*

㉕ *Because it is so verdant and green, as a result of frequent rainfalls, mild climate, and rich soil.*

㉖ *Yucatan.*

㉗ *Des Moines.*

㉘ *The Lone Star State.*

㉙ 1. *Arkansas, Idaho, Illinois, Kentucky, Missouri, Ohio, Tennessee, and Wisconsin.* 2. *Alabama, Arkansas, Indiana, Kentucky, Ohio, and Oregon.* 3. *California, Kentucky, Indiana, Michigan, Mississippi, New Hampshire, North Carolina, Pennsylvania, and Virginia.* 4. *Alabama, Iowa, Kentucky, Maine, Nebraska, and New Mexico.* 5. *Idaho, Indiana, Kansas, Kentucky, Michigan, Pennsylvania, and Vermont.* 6. *Illinois, Indiana, Iowa, Louisiana, Maine, Maryland, New Hampshire, New York, and Ohio.* 7. *New York.* 8. *Missouri, New York, and Ohio.* 9. *Georgia, Illinois, Indiana, Maine, Mississippi, New York, Oregon, Pennsylvania, Tennessee, and Wisconsin.* 10. *New York.*

㉚ 1. *Kentucky, Michigan, Missouri, and Pennsylvania.* 2. *Pennsylvania.* 3. *New York.* 4. *New Mexico, New York, and Ohio.* 5. *New York.* 6. *Illinois.* 7. *Illinois, Michigan, Ohio, and Wisconsin.*

㉛ *Kuwait.*

㉜ *The Red Sea.*

㉝ *The Black Sea.*

㉞ *The Yellow Sea.*

㉟ *The Yellow River.*

㊱ *The Blue Ridge Mountains. The highest peaks are not in Virginia, though; they are in North Carolina.*

㊲ *Rainbow Bridge.*

㊳ *(a) King Charles the First, (b) Charleston, originally called Charlestown.*

㊴ *Connecticut.*

㊵ *"Holy Faith."*

㊶ *Turin.*

# Countries and Continents

It would seem reasonable to expect that a geography book purchased today would remain timely and accurate for another four or five years. Unfortunately, such is not the case. Our planet is constantly in turmoil, geographically speaking, necessitating significant revisions in all kinds of reference books year after year. You would expect that countries, at least, might remain pretty much the same for years on end. But, like the large corporations listed on Wall Street, they are constantly merging, going out of business, being formed or reformed, changing names, and most certainly going through political upheavals.

*How is your knowledge of the countries of the world? Let's find out by answering the questions in this chapter.*

**❶** How many countries are there in the world today: (a) 88, (b) 125, or (c) more than 160?

**❷** How many continents are there and can you name all of them?

**❸** What is the oldest country in the world? (Hint: It lies in Asia and had a different name until a little over half a century ago.)

**❹** How many countries have no seacoast: (a) 12, (b) 18, or (c) 26?

**❺** China is slightly larger than the United States, stretching far enough from east to west so that one might be in bright sunlight while the other is in darkness. How many time zones would you go through if you were on a flight from Shanghai on the east coast to Urumchi, a city that lies in the far northwest?

**❻** Egypt is an ancient land, ruled by a succession of conquerors. When the Arabs invaded Egypt in the seventh century, they built new cities and towns, including the present capital of the country and the largest city in Africa. Can you name it?

**7** Are there more countries above the equator or below it? Or are the numbers about comparable?

**8** Which country has the most coastline?

**9** What country is composed of more ethnic groups than any other nation?

**10** Assyria was once a powerful nation which, by the year 700 B.C., had conquered all of Syria, Israel, and Egypt before gradually losing control and territory. Is Assyria still a country and, if so, where is it?

**11** Chile is one of the mostly oddly shaped of the world's nations, stretching 3,000 miles from north to south, but less than 200 miles wide for most of its length. For a two-parter: (a) where is Chile located, and (b) what countries border it?

**12** Can you name the country that was once known as Abyssinia and whose capital, founded in 1887, is Addis Ababa ("New Flower")?

**13** What is the difference between the Republic of China and the People's Republic of China?

**14** In 1816 the American Colonization Society was founded. Its goal was to free slaves in the United States and send them back to Africa, to a new nation established for that purpose. For a two-parter: (a) what is the name of this made-to-order country, and (b) what is its capital?

**15** New Caledonia is an island in the Pacific Ocean. It is a tropical haven in the South Seas noted for its warm climate. *Old* Caledonia is completely different in locale and environment. What is it?

**16** Tierra del Fuego is a remote, forbidding shore that was named by Magellan as he rounded Cape Horn on a voyage of discovery in the early sixteenth century. Tierra del Fuego ("Land of Fire") and Cape Horn lie at the southern tip of which continent?

**17** This small, mountainous country, near the southwestern region of Yugoslavia, borders on Greece to the south. Name it.

**18** For a two-part question: (a) name the six island countries of the West Indies, and (b) of these six countries, which one is farthest west and which is farthest east?

**19** Where is Puerto Rico? What is its political standing as a government?

**20** Which countries comprise the region we refer to as Central America?

**21** Ecuador is a small country on the edge of the Pacific Ocean in South America, named because of its position right on the equator. What is its capital?

**22** This country in the northwest corner of South America has earned a dubious reputation as a source of illegal drugs. For years, however, it has been known as a major producer of coffee. Name the country.

**23** Two countries in South America have no navy, for the simple reason that they have no ocean coastline. What are they?

**24** One country in Central America (a region often plagued by civil wars) is unique in that it has no standing army. What country could be so brave in the midst of military actions on all sides?

**25** This country, second in size in South America only to Brazil, is a major farming area, basing much of its economy on crops and livestock. For this reason, it relies heavily on wide, flat, grassy plains with rich soil called *pampas*. Name the country.

**26** One of the smallest countries in South America was settled largely by Europeans and is so cosmopolitan in nature that forty percent of the people live in the capital city, Montevideo. What is this country?

**27** Located in western Europe, this federal republic was once the province of Helvetia. For a two-parter: (a) name the country, and (b) name the countries that border it.

**28** Approximately how many countries are there in the continent of Asia: (a) 17, (b) 24, (c) 39, or (d) 61?

**29** At the end of World War II, Korea (a country whose roots were largely Japanese and Chinese) was divided into two sections: North Korea and South Korea. Name the capitals of each.

**30** Known as Europe's oldest existing state, the world's smallest republic has a population of less than 20,000 people, most of whom speak Italian. Can you name this tiny land?

**31** Kashmir is one of the most beautiful regions of the East, covered with lofty, rugged mountains and split by lush, magnificent valleys, including the famed Vale of Kashmir, long the subject of song and story. Where is Kashmir?

**32** Southwest Asia contains countries that have constantly been in the news, including Afghanistan, Iran, Iraq, Kuwait, and Saudi Arabia. Answer the following questions pertaining to these countries:

(a) Which of these countries border on Iraq?

(b) Which country was invaded by Russia in 1979?

(c) Which countries have shorelines on the Persian Gulf?

(d) Which of these countries produce oil?

(e) Which of these countries are republics?

(f) In which countries is Arabic the major language?

**33** The above-mentioned countries of Southwest Asia have a neighbor that is little known to Westerners and is a poor country, with no oil. Bordered on the north and east by Saudi Arabia, it lies in a mountainous area on the southern-most shore of the Red Sea. Can you name it?

**34** The United Arab Emirates, formerly called the Trucial States, consist of seven tiny countries, of which Abu Dhabi and Dubai are the largest. Where is this group of nations located?

**35** Ironically, war-torn Lebanon, and particularly the shattered city of Beirut, was once an important tourist mecca. Can you explain why?

**36** This country is unique in that it is situated on two continents, Asia and Europe. The dividing line is the Bosporus, a strait that connects the Sea of Marmara to the Black Sea. Name the country.

**37** Five nations in northern Africa have shorelines on the Mediterranean. Can you name them?

**38** Which of the five nations mentioned in the preceding question is the largest? Which is the smallest?

**39** West Africa is the term used to designate the cluster of twelve countries on the Atlantic side of the continent. How many can you name?

**40** Freetown is the capital of a country founded by the British as a home for former slaves. Name the country.

**41** The largest of the eight countries that are classified as being Central Africa is Zaire, which is almost as large as all the others put together and two-thirds the size of Europe. What was its former name?

**42** To the southeast of Kenya in the Indian Ocean lies an island slightly smaller than Texas in size and the home of some eleven million people who are largely Malayan-Indonesian. What is the name of this nation?

**43** Southern Africa is comprised of six nations, of which the Republic of South Africa is the largest and by far the most prosperous. The two smallest countries in southern Africa lie entirely within the borders of the largest country. Can you name either one of them? (Hint: One was formerly called Basutoland and the other is referred to as "the Switzerland of Africa" because of its high, magnificent mountain scenery.)

**44** South Africa is unique in that it has not one capital, but three. Can you name at least two of the three?

**45** What is the capital of the largest country in Central America?

**46** What countries and bodies of water form the borders of Costa Rica?

**47** What is the largest country in the world today in terms of area, about two-and-one-half times the size of the United States?

**48** What countries are grouped under the heading "Scandinavia"?

**49** For a two-parter: (a) which Scandinavian country is the largest, and (b) what is its capital?

**50** France has three coast-lines. Name the bodies of water that comprise them.

**51** The Pyrenees Mountains form a distinct natural border between what two countries?

**52** Match these European countries with their capitals:

| Country | Capital |
|---|---|
| **1.** Liechtenstein | **(a)** Belgrade |
| **2.** Bulgaria | **(b)** Copenhagen |
| **3.** Belgium | **(c)** Athens |
| **4.** Portugal | **(d)** Vaduz |
| **5.** Denmark | **(e)** Lisbon |
| **6.** Romania | **(f)** Sofia |
| **7.** Greece | **(g)** Brussels |
| **8.** Yugoslavia | **(h)** Bucharest |

**53** Of the three major countries in Eastern Europe, name: (a) the northernmost, and (b) the southernmost.

**54** What European country has so steadfastly maintained its neutrality with neighboring nations that it has not been involved in a foreign war since 1515?

**55** Referring to the country in the preceding question: (a) name its capital, and (b) name its largest city.

**56** Which continent has a coastline that is longer and more irregular than that of any other continent?

**57** This is the largest country in Europe, with the exception of the Soviet Union. (a) Name it, and (b) name the countries it borders.

**58** Next to its capital, what is the second largest city in the country mentioned in the preceding question?

**59** In terms of land size, what is the second largest country in the world?

**60** One of the most popular regions of Spain, for Spaniards and foreigners alike, is the Costa del Sol. What is the Costa del Sol and where is it? For extra credit, what does the name mean in English?

**61** High in the eastern Pyrenees between Spain and France lies a tiny independent principality, which depends primarily on sheep raising and agriculture. What is the name of this country, which has only 185 square miles, half the size of New York City.

**62** The southwestern tip of the continent of Europe is Cape St. Vincent, near the port of Lagos, from which Prince Henry the Navigator planned expeditions to the New

World during the Age of Discovery. In what country are this cape and city located?

**63** This European country is long (1,100 miles) and narrow, forming a mountainous strip along the North Sea, its seacoast sharply serrated with hundreds of inlets and deep fjords and fringed with small islands. Name the country.

**64** What is the name of the capital and largest city of the country in the preceding question?

**65** Most of Scotland is rural, its people relying on agriculture and production from the sea. But its largest city is known as "Britain's greatest industrial center." What city is that?

**66** Except for Antarctica, which continent has the fewest major ports?

**67** The people who live in Buenos Aires are known as Portenos ("people of the port.") Of what country is Buenos Aires the capital?

## Answers

**1** *(c) There are more than 160 countries (the number changes from time to time), plus more than 60 lands that are territories.*

② *The seven continents of the world are Africa, Antarctica, Asia, Australia, Europe, North America, and South America.*

③ *Iran, which was known as Persia until 1934. It has been independent since 529* B.C.

④ *(c) Twenty-six countries have no seacoast at all.*

⑤ *By all rights, China should have as many time zones as the United States. Yet it has only one, to which all of the country's more than one billion people have to adjust. Although this seems impractical to Westerners, that is the system the government in Beijing prefers.*

⑥ *Cairo.*

⑦ *There are more than three times as many countries above the equator as below it.*

⑧ *Canada, with about 155,000 miles of coastline, or about six times more than that of Australia.*

⑨ *The USSR, which has some 280 million people and in which more than one hundred languages are spoken.*

⑩ *Assyria is no longer a country. It was eventually wiped out by its enemies.*

⑪ *(a) Chile lies along the southwestern coast of South America on the Pacific*

*Ocean. (b) It is bordered by Peru to the north, and Bolivia and Argentina to the east.*

⑫ *Ethiopia.*

⑬ *The People's Republic of China is the old mainland China. The Republic of China is Taiwan, formerly Formosa, which broke off from mainland China in 1949 and lies entirely on an island to the south.*

⑭ *(a) Liberia. (b) The capital, Monrovia, was named after James Monroe, then president of the United States.*

⑮ *The original Caledonia is Scotland, an ancient name that is now used only in poetry and mythology.*

⑯ *South America.*

⑰ *Albania.*

⑱ *(a) The Bahamas, Barbados, Cuba, the Dominican Republic, Haiti, and Jamaica. (b) Cuba is farthest west and Barbados farthest east.*

⑲ *Puerto Rico lies in the middle of the chain of islands that make up the West Indies. It is a self-governing commonwealth of the United States.*

⑳ *In order, from north to south, the nations of Central America include Belize (British Honduras), Guatemala, Honduras, El Sal-*

vador, *Nicaragua, Costa Rica, and Panama. From a geological standpoint, the southern part of Mexico is also included.*

21 *Quito is the capital of Ecuador. It is unique because it is situated at 9,350 feet above sea level where, despite its position on the equator, it enjoys an average temperature of 55°F.*

22 *Colombia.*

23 *Bolivia and Paraguay, neighboring countries high in the Andes, have no coastline.*

24 *Costa Rica, known as the most peaceful of all the countries of Latin America, abolished its army many years ago. When threatened, this tiny nation can quickly organize its citizens into a volunteer defense force.*

25 *Argentina. The* Pampa *of central and northern Argentina covers 250,000 square miles. Reminiscent of scenes from the Old West are the cowboys, called* gauchos, *who ride herd.*

26 *Uruguay, which is tucked into a 68,000-square-mile pocket on the Atlantic side of the continent.*

27 *(a) Switzerland. (b) Germany, Austria, Liechtenstein, Italy, and France.*

28 *(c) At last count, Asia consisted of thirty-nine independent countries, dominated in size by China and Russia.*

㉙ *Pyongyang is the capital of North Korea. Seoul is the capital of South Korea.*

㉚ *San Marino, which lies in the Apennine Mountains near the Adriatic Sea, southwest of Rimini in northern Italy. Its origins as an independent state go back to the fourth century.*

㉛ *Kashmir is a predominantly Moslem region in northwestern India. It has long been the subject of dispute between India, which controls two-thirds of the area, and Pakistan, which controls one-third.*

㉜ *(a) Iran, Kuwait, and Saudi Arabia. (b) Afghanistan. (c) Iran, Iraq, Kuwait, and Saudi Arabia. (d) Iran, Iraq, Kuwait, and Saudi Arabia. (e) Iran, Iraq and Afghanistan. (f) Iraq, Saudi Arabia, and Kuwait.*

㉝ *Yemen. It was once part of the ancient kingdom of Sheba.*

㉞ *The United Arab Emirates (UAE) lies in a crescent-shaped area at the southern part of the Arabian Gulf and to the east of Saudi Arabia.*

㉟ *Lebanon is situated on the eastern shores of the Mediterranean, and has a mountain range over 10,000 feet high whose snow-covered slopes are ideal for skiing. For much of the year, the country enjoys a warm, sunny climate that attracts summer tourists as well as skiers.*

**㊱** *Turkey.*

**㊲** *From west to east, they are Morocco, Algeria, Tunisia, Libya, and Egypt.*

**㊳** *Algeria is the largest with an area of 520,000 square miles. Tunisia is the smallest, with 63,000 square miles.*

**㊴** *They include (from west to east): Senegal, Gambia, Guinea-Bissau, Guinea, Sierra Leone, Liberia, the Ivory Coast, Ghana, Togo, Benin (formerly Dahomey), and Nigeria.*

**㊵** *Freetown is the capital of Sierra Leone.*

**㊶** *The Belgian Congo.*

**㊷** *Madagascar, which is also known as the Malagasy Republic.*

**㊸** *One is Lesotho (formerly Basutoland), and the other is Swaziland, a kingdom slightly smaller than New Jersey.*

**㊹** *Cape Town (legislative), Pretoria (administrative), and Bloemfontein (judicial).*

**㊺** *Managua, the capital of Nicaragua, which is about the size of the state of Iowa.*

(46) *Nicaragua to the north, the Pacific Ocean to the west, the Caribbean Sea to the east, and Panama to the south.*

(47) *The USSR, which occupies about one-sixth of the land mass of the globe.*

(48) *Norway, Sweden, Finland, Denmark, and Iceland.*

(49) *(a) Sweden is the largest. (b) Its capital is Stockholm.*

(50) *The Atlantic Ocean to the west, the Mediterranean Sea to the south, and the English Channel to the north.*

(51) *Spain and Portugal.*

(52) *1(d) Liechtenstein/Vaduz, 2(f) Bulgaria/Sofia, 3(g) Belgium/Brussels, 4(e) Portugal/Lisbon, 5(b) Denmark/ Copenhagen, 6(h) Romania/Bucharest, 7(c) Greece/Athens, and 8(a) Yugoslavia/Belgrade.*

(53) *(a) Poland is the northernmost, on the Baltic Sea. (b) Hungary is southernmost. Czechoslovakia is sandwiched in between.*

(54) *Switzerland*

(55) *(a) Bern; (b) Zurich.*

(56) *Europe. From any point in Europe (except for parts of central Russia)*

*it is impossible to be more than 400 miles from the sea.*

⑰ *(a) France. (b) Belgium, Luxembourg, Spain, Germany, Switzerland, and Italy.*

㊽ *Marseilles.*

㊾ *Canada, which covers some 3,852,000 miles.*

⑳ *The Costa Del Sol lies along the southern part of Spain on the Mediterranean. Its name means "coast of the sun."*

㉑ *Andorra.*

㉒ *Portugal.*

㉓ *Norway.*

㉔ *Oslo.*

㉕ *Glasgow.*

㉖ *Australia.*

㉗ *Argentina.*

# A Tour of the United States

# H

*aving already made a world tour, let's narrow our geographical target and take an instructive tour of the United States. You might be surprised at the extent and variety of things you can see and do within America's borders.*

*Let's start the day early with the first question.*

**1** You have gone on a camping and hiking trip along part of the New England coast and are standing at dawn on the very place where the sun first strikes the United States each day (when it is not overcast, of course). Where are you?

**2** If you were to sail up the Chesapeake Bay and land once on each of the states that it touches, how many such landings would you make?

**3** Here's an easy one: You are visiting the two states that border on no other states in the United States. Which ones are they?

**4** Another easy one: You are visiting a state that borders on only one other state. Which one is that?

**5** If you were at a point in the United States where you could easily toss four stones and have them land in four different states, where would you be?

**6** Where would you go if you wanted to spend a summer vacation motoring through the Twin States?

**7** Sailing down the Mississippi River from its origin in Lake Itasca near the Canadian border to its mouth in the Gulf of Mexico, you would pass by a dozen or more major cities in ten states. Only one of these is a state capital. Can you name it?

**8** You find yourself at the most desolate end of the United States, in a community of little more than 2,000 people. It is noted for being the northernmost town in the United States. Where are you?

**9** You are visiting New York City and someone asks you what European seafarer first sailed in the waters that surround

Manhattan. What would you answer? (Hint: He has a bridge named after him and it was *not* Henry Hudson.)

**10** As part of a vacation trip, you have decided to visit the Grand Canyon. Thinking it might be fun to visit other wonders in the same state, you plan to write to the state tourist board for information. Where would you write?

**11** On a trip that takes you through Oklahoma City, you are amazed at what you see right in the heart of town. What is so unique about this place?

**12** Considering the continental United States (and thus excluding Alaska and Hawaii), where would you go to be at these extremes: (a) the northernmost point, (b) the southernmost point, (c) the point farthest west, and (d) the point farthest east?

**13** Remaining within the borders of the United States, where would you drive to cover the shortest distance from the Atlantic to the Pacific?

**14** If you wanted to stand on the exact geographic center of the continental United States, where would you go?

**15** What state has the *lowest* average elevation?

**16** What state has the *highest* average elevation?

**17** Can you name the six New England states?

**18** Together, the Plateau states constitute about one quarter of the area of the United States. Can you name these eight states?

**19** More states touch Tennessee than any other in the nation. Can you name the eight states that border it?

**20** Texas got its name when native Indians greeted the first "tourists," Spanish explorers, with cries of *Techas!* ("Friends!") and indicated their peaceful nature and neighborly intentions. Although Houston is the largest and best known city in the state, it is not the capital. What city is?

**21** Match the following states with their nicknames:

1. Connecticut        (a) Palmetto State
2. Pennsylvania       (b) Peach State
3. New Hampshire      (c) Old Dominion
4. Georgia            (d) Keystone State
5. South Carolina     (e) Volunteer State
6. Missouri           (f) Granite State
7. Tennessee          (g) Prairie State
8. Virginia           (h) Sunshine State
9. Illinois           (i) Nutmeg State
10. Florida           (j) Show Me State

**㉒** The smallest and easternmost of the Great Lakes is Ontario, which means "Beautiful," a name adopted also by the Canadian province of Ontario. Can you name two or more of the six states of the United States that lie directly south of this province?

**㉓** If you were determined to visit every major national park in the United States, how many stops would you have to make: (a) 17, (b) 26, (c) 35, (d) 45?

**㉔** You do not have to leave the continental borders of the United States to see real deserts. There are four that rank among the twenty-five largest in the world. Can you name them?

**㉕** The United States has four of the fifteen highest waterfalls in the world. They are all located in a single national park. Name it.

**㉖** Where would you go if you wanted to visit: (a) the Cotton Belt, and (b) the Corn Belt?

**㉗** The itinerary for a tour of the West includes the Great Basin. (a) Where is this, and (b) what is this?

**㉘** A visit to Kauai would be interesting because it is the only place in the United States where you could see all of these geological wonders: A "wailing" cave, sands that "bark" when you walk on them, spouting horns that shoot jets of sea water into the air, bogs that tremble and quiver, and a river that seems to run uphill. What is Kauai and where is it?

**㉙** Match the following national parks with the states in which they are located. For extra credit, describe a distinguishing feature of each one.

| **Park** | **State** |
| --- | --- |
| **1.** Acadia | **(a)** California |
| **2.** Badlands | **(b)** Colorado |
| **3.** Isle Royale | **(c)** Utah |
| **4.** Mesa Verde | **(d)** South Dakota |
| **5.** Sequoia | **(e)** Virginia |
| **6.** Shenandoah | **(f)** Michigan |
| **7.** Wrangell-Saint Elias | **(g)** Maine |
| **8.** Zion | **(h)** Alaska |

**㉚** In order, which states would you pass through if you were driving south through the Middle Atlantic States?

**㉛** If you were to take a barge trip on the Erie Canal, where would you travel?

**㉜** You want to visit the homes of these great American presidents, which are today open to tourists: (a) Franklin Delano Roosevelt, (b) Andrew Jackson, (c) Thomas Jefferson, and (d) Theodore Roosevelt. Name the home of each president, and give the location.

㉝ In the United States, the following nine places have been designated "National Seashores." Match the names with the locations:

| Seashore | State |
|---|---|
| **1.** Asateague Island | **(a)** Georgia |
| **2.** Cape Cod | **(b)** New York |
| **3.** Cape Hatteras | **(c)** Florida, Mississippi |
| **4.** Cape Lookout | **(d)** Massachusetts |
| **5.** Cumberland Island | **(e)** Texas |
| **6.** Fire Island | **(f)** North Carolina |
| **7.** Gulf Islands | **(g)** Maryland and Virginia |
| **8.** Padre Island | **(h)** California |
| **9.** Point Reyes | **(i)** North Carolina |

㉞ You are going to fly across the country, stopping off at various cities to see the sights. But you sacrifice one major stop-over in order to avoid the busiest airport in the country. Which one is that?

㉟ What state can you walk to from Nevada across the Hoover Dam?

㊱ You don't have to travel abroad to go boating on the following rivers. Where are these located in the United States: (a) The Thames (England), (b) Yellow River (China), (c) Jordan (Lebanon/Jordan), (d) Mackenzie (Canada), (e) Peace (British Columbia), and (f) Tiber (Italy)?

**37** If you were to visit "The Breadbasket of America," what state would you be in?

**38** Which states have shorelines on the Gulf of Mexico?

**39** Which state is seventy-five percent surrounded by three lakes?

**40** There are four states that have geographic features known as panhandles. For this two-part question: (a) what is a panhandle, and (b) where are these located?

**41** If you wanted to see the La Brea Pits, where would you go? For extra credit, what are they?

**42** The place with the most rainy days in the year happens to be in the United States. Can you name it?

**43** Skiing has become very popular in the United States in the past few decades. Name the state in which you can find the following ski resorts: Aspen, Steamboat Springs, and Vail.

**44** This small island, about twelve miles long, lies off the southernmost tip of South Carolina, is shaped like a boot, and has become increasingly popular with tourists during the past decade. Name it.

**45** The Ozark Plateau, commonly referred to as "the Ozarks," have forests, streams, and many other scenic beauties. Among the other attractions, too, are several large man-made lakes that were formed by dams across the White and Black Rivers. Where are the Ozarks?

**46** You are planning to take the beautiful seventeen-mile scenic drive along the Monterey Peninsula. Where is Monterey?

**47** You are a Civil War history buff and would like to visit Fort Sumter, renowned as the scene of the first engagement of the Civil War. Where is Fort Sumter? For extra credit, where did its name come from?

**48** Twenty-four time zones were established in the world in the 1880s to coincide approximately with meridians at successive hours from the observatory at Greenwich, England. There are four time zones in the United States. What are they named, in order, from east to west?

**49** Referring to the preceding question, in what time zones are the following United States cities: (a) Las Vegas, Nevada,

(b) Washington, D.C., (c) Nashville, Tennessee, (d) Phoenix, Arizona, and (e) Houston, Texas?

**50** Maine sits at the edge of a fifth time zone that runs along its eastern border and affects the easternmost Canadian provinces. What is this called?

**51** Tens of thousands of sandhill cranes can be seen every March in the San Luis Valley on their annual trek from Texas to Canada. The valley is located just south of the Rio Grande National Forest and east of the San Juan National Forest. To what state would you have to go to observe this annual migration?

**52** In the early sixteenth century, Spanish explorers in the American west founded a town they called Don Fernando de Taos, now simply called Taos. It was the site of an Indian pueblo established there almost one thousand years ago and partially reconstructed as a tourist site in recent times. In what state is Taos and its Indian pueblo?

**53** You would like to visit Lake Pontchartrain, one of the largest lakes in the country. A city that is a major tourist center sits on this lake, which covers some 630 square miles and has the longest multispan bridge in the world. What is the name of this popular city?

**54** The District of Columbia was originally named in honor of Columbus

when it was established by congressional acts of
1790 and 1791. The District was later created as a
municipal corporation in 1871. Besides Washington, what other town is part of the District?

**55** You want to visit all
the states that comprise the Midwest. Can you
name them?

# Answers

① *At the top of Mount
Katahdin, Maine, which rises to 5,267 feet and is
the highest peak in the state.*

② *You would land only
twice: on the shores of Maryland and Virginia. You
could stretch it a bit, though, by sailing a short way
up the Susquehanna River where it empties into the
Chesapeake, and thus land on Pennsylvania shores.*

③ *Hawaii and Alaska.*

④ *There is only one such
state. That is Maine; it borders on New Hampshire.*

⑤ *At a place called Four
Corners where these states meet: Utah, Colorado,
Arizona, and New Mexico.*

⑥ *New Hampshire and
Vermont.*

⑦ *St. Paul, Minnesota, which sits on bluffs opposite its sister city, Minneapolis.*

⑧ *Barrow, Alaska, where the population consists mainly of eskimos who hunt and fish and endure long winters with temperatures often more than 50°F below zero.*

⑨ *Giovanni da Verrazano, an Italian explorer, who sailed into what is now New York Bay in 1524. The Verrazano-Narrows Bridge, linking Brooklyn and Staten Island, was named after him.*

⑩ *Phoenix , Arizona.*

⑪ *Oil rigs operate there.*

⑫ *(a) The northern border of Minnesota, (b) Cape Sable, Florida, (c) Cape Alava, Washington, and (d) West Quoddy Head, Maine.*

⑬ *The shortest distance from ocean to ocean is a straight line drawn from a point near Charleston, South Carolina, to a point near San Diego, California. The distance is 2,152 miles as the crow flies.*

⑭ *To a point seventeen miles due west of Castle Rock, South Dakota at latitude 44° 58' N and longitude 103° 46' W.*

⑮ *Delaware. Its surface, if leveled, would be only sixty feet above the sea.*

⑯ *Colorado. Its* average *elevation is a breathtaking 6,800 feet.*

⑰ *Connecticut, Maine, Massachusetts, New Hampshire, Rhode Island, and Vermont.*

⑱ *Arizona, Colorado, Idaho, Missouri, Nebraska, New Mexico, Utah, and Wyoming. Their total area is about one quarter of the United States.*

⑲ *In geographical order clockwise around Tennessee, they are Kentucky, Virginia, North Carolina, Georgia, Alabama, Mississippi, Arkansas, and Missouri.*

⑳ *Austin, named after a Texas colonizer, Stephen Fuller Austin.*

㉑ *1(i) Connecticut/Nutmeg State, 2(d) Pennsylvania/Keystone State, 3(f) New Hampshire/Granite State, 4b) Georgia/Peach State, 5(a) South Carolina/Palmetto State, 6(j) Missouri/Show Me State, 7(e) Tennessee/Volunteer State, 8(c) Virginia/Old Dominion, 9(g) Illinois/ Prairie State, and 10(h) Florida/Sunshine State.*

㉒ *From west to east in order: Minnesota, Wisconsin, Michigan, Ohio, Pennsylvania, and New York.*

㉓ *(c) There are thirty-five such parks.*

㉔ *In order of size they are: the Mojave, California, 15,000 square miles; the Painted Desert, Arizona, 5,000 square miles; Great Salt Lake, Utah, 4,000 square miles; and Death Valley, California/Nevada, 1,500 square miles.*

㉕ *Yosemite National Park in east central California, which has two other waterfalls in the world's top twenty.*

㉖ *(a) Alabama, Georgia, and Mississippi. (b) Iowa, Illinois, and Indiana.*

㉗ *(a) The Great Basin lies mostly in Nevada, but extends into the surrounding states of California, Oregon, Idaho, and Utah. (b) It is a desert region containing numerous fault-block mountains that rise to heights of more than 10,000 feet.*

㉘ *Kauai is an island, the northernmost one in Hawaii.*

㉙ *1(g) Acadia/Maine/granite mountains and scenic seacoast, 2(d) Badlands/South Dakota/fossils and rugged, eroded slopes, 3(f) Isle Royale/Michigan/heavily forested islands, 4(b) Mesa Verde/Colorado/prehistoric cliff dwellings and pueblo houses, 5(a) Sequoia/California/stands of giant sequoia trees reaching 300 feet and more, 6(e) Shenandoah/Virginia/skyline drive along the crest of the Blue Ridge Mountains, 7(h) Wrangell-Saint Elias/Alaska/glaciers and high peaks, 8(c) Zion/Utah/colorful canyons and pictur-esque sandstone cliffs.*

③⓪ *New York, New Jersey, and Pennsylvania.*

③① *Between Albany, New York, and Buffalo, New York. This 360-mile canal connects the Hudson River with the Great Lakes.*

③② *(a) F. Roosevelt: Hyde Park; Hyde Park, New York. (b) A. Jackson: The Hermitage; Nashville, Tennessee. (c) T. Jefferson: Monticello; Charlottesville, Virginia. (d) T. Roose-velt: Sagamore Hill; Oyster Bay, New York.*

③③ *1(g) Asateague Island/ Maryland and Virginia, 2(d) Cape Cod/Massachu-setts, 3(i) Cape Hatteras/North Carolina, 4(f) Cape Lookout/North Carolina, 5(a) Cumberland Island/ Georgia, 6(b) Fire Island/New York, 7(c) Gulf Is-lands/Florida and Mississippi, 8e) Padre Island/ Texas, and 9(h) Point Reyes/California.*

③④ *O'Hare Airport in Chi-cago, which has the dubious distinction of being the busiest in the world, with a takeoff or landing about every forty-eight seconds, around the clock.*

③⑤ *Arizona.*

③⑥ *(a) Connecticut, (b) Wisconsin, (c) Utah, (d) Oregon, (e) Florida, and (f) Montana.*

(37) *Kansas, which produces more wheat than any other state.*

(38) *In order, from east to west, they are: Florida, Alabama, Mississippi, Louisiana, and Texas.*

(39) *Michigan, which is bordered by Lake Michigan, Lake Huron, and Lake Erie on three sides.*

(40) *(a) A panhandle is a strip of land projecting from the main body of an area and shaped like the handle of a pan. (b) Alaska, Florida, Texas, and West Virginia.*

(41) *The La Brea Tar Pits are located in Los Angeles, California. (b) They are asphalt pits in which extensive remains of prehistoric animals and plants have been found.*

(42) *Mount Waialeale in Hawaii on the island of Kauai. In the course of an average year, there are only fifty days when you could visit this 5,080-foot peak without getting drenched.*

(43) *Aspen, Steamboat Springs, and Vail are all in Colorado.*

(44) *Hilton Head.*

(45) *The Ozarks lie chiefly in southern Missouri and northern Arkansas, and partly in Oklahoma and Kansas.*

㊻ *Monterey juts out into the Pacific Ocean halfway up the California coast, south of San Jose and San Francisco.*

㊼ *Fort Sumter lies off Charleston, South Carolina on a small island. It was named after General Thomas Sumter.*

㊽ *Eastern Time Zone, Central Time Zone, Mountain Time Zone, and Pacific Time Zone.*

㊾ *(a) Las Vegas/Pacific, (b) Washington/Eastern, (c) Nashville/Central, (d) Phoenix/Mountain, and (e) Houston/Central.*

㊿ *Atlantic Time Zone.*

(51) *Colorado.*

(52) *New Mexico.*

(53) *New Orleans, Louisiana, which lies between the lake and the Mississippi River.*

(54) *Georgetown. Together, they comprise Washington County.*

(55) *Ohio, Indiana, Illinois, Michigan, Wisconsin, Minnesota, Iowa, Missouri, Kansas, and Nebraska.*

# A Sampling
# of Geology

Since geography and geology are two disciplines that overlap and share many subject areas, it is worthwhile to consider some of the facts and factors that have a common relationship. Geography is the science of place, the study of the surface of the earth, the location and distribution of its physical and cultural features and the interrelation of these features as they affect humans. Geology is the science of the earth's history, composition, resources, and structure, and the associated processes.

What has been seen yesterday in any given place is not necessarily what will be seen there tomorrow. That is part of the excitement of geology and the reason *why cartographers will never run out of changes to make on their maps.*

*How much do you know about geology? Let's start with the basics.*

**❶** In size, the earth is in the middle position among the nine planets in the solar system. What is its diameter: (a) 3,000 miles, (b) 4,220 miles, (c) 7,926 miles, or (d) 9,567 miles?

**❷** What is the term for a plain (such as ones near the Rio Grande River in New Mexico) that has broad, gently sloping surfaces constructed from stream-deposited sediment?

**❸** A national monument located along the White River in South Dakota is the world's best and most extensive example of a curious geological formation. For this two-parter: (a) name the monument, and (b) give the term used to describe this type of formation.

**❹** The Gulf Stream is well marked on charts and maps and is of great interest because its warm waters have decided effects on the climate of islands and mainland shores. Where does the Gulf Stream start and end?

**❺** The topography of the land depends a great deal upon the geologic foundations that lie beneath it. What do you think is the most predominant rock underlying the continental United States?

**6** Once an immense prehistoric lake that was expanded during the Ice Age in North America and covered almost 20,000 square miles, this place has long been a laboratory for the study of its origins, demise, and present condition. In a less scientific way, it has also served as a testing ground for super racing cars. Name this former lake and its location.

**7** Geography books contain illustrations showing regions few people ever see but that are vital to the nature and profile of our planet. One example is the gently sloping sea bed that fringes each of the continents and varies in width from only a few miles to 250 miles and in depth from the low-tide mark to about 600 feet. What is this formation called?

**8** The coral reefs and atolls of the Pacific Ocean provide dramatic evidence of the way in which minute organisms (in this case, coral) can, over millions of years, build up islands large enough and solid enough for human habitation. During what war were these geologic wonders strategically vital? Why?

**9** The North Sea is a geologist's paradise because of formations lying beneath the sea bed that are of significance in the world's economy. Can you explain why?

**10** Hudson Bay, whose northern reaches are on the Arctic Circle, provides valuable data about the nature and movement of glaciers during the Ice Age. One example is the existence of a shoreline 840 feet above sea level, providing evidence that there was an immense uplift in the land following the retreat of the last ice sheet. Here's a two-parter: (a) what kind of geographical feature is Hudson Bay, and (b) where is it located?

**11** According to a much-accepted geological theory, North America and Europe were once joined as a single super-continent that existed 225 million years ago, hypothetically called *Laurasia*. What is the term used to explain the separation that took place in prehistoric times?

**12** Mountains referred to as "sugar loaf" types (because they resemble conical loaves of hard, refined sugar) are familiar features in a number of lands around the globe, often studied to determine their origins and the geological nature of their curious shape. The most familiar and certainly most visited is Sugarloaf Mountain, which rises about 1,300 feet and looks out over a magnificent harbor, frequently called the most beautiful in the world. Where is this unique peak?

**⑬** Papua New Guinea is frequently visited by geologists interested in studying the layered rock formations that make up much of the surface but in prehistoric times lay six miles beneath the Pacific Ocean. The mountainous nature of the land is surprising in light of the fact that the nearest continent, which lies just to the south, is noted for its lack of high mountains. What is the continent so close to it?

**⑭** In what kind of geological formation would you find rocks that look as though they might be as heavy as chunks of granite but are as light as balsa wood?

**⑮** There is a cave in the Black Hills of South Dakota that is unique because strong air currents constantly blow through its 10.5 miles of explored passageways, the direction and force depending upon the pressure of the air outside. What is the name of this cave?

**⑯** Iceland lies almost on the Arctic Circle in the Norwegian Sea. Despite its northern location, it has some contradictory thermal features that make it a geological wonder. What are these features?

**⑰** The southern coast of Portugal is characterized by steep cliffs reminiscent of the Amalfi coast of Italy, and by strange volcanic formations and sea-sculptured rocks that look like something from another planet. What is this coast named?

**18** Lake Maracaibo on the northwest coast of Venezuela has probably been visited by more American geologists than any other place of like size in South America. One look at the lake will tell you why this is so. What would you see if you looked at the lake?

**19** A famous body of water that lies along the Jordan-Israel border and whose surface is 1,300 feet below sea level is aptly named the Dead Sea. Why?

**20** Wherever we live or travel, we are likely to be dependent upon an underground formation that, because of good porosity and permeability, is able to transmit water in sufficient quantities to supply springs and wells. Is this formation called: (a) an aquifer, (b) an artesian well, or (c) lithification?

**21** The Pleistocene Epoch was a geological period that had a great deal to do with shaping the surface of the earth as we know it today. What is another name for this epoch and why was it so significant?

**22** A seamount is a term for a well-defined formation that is found in the oceans, but not easily studied. What do you think it is: (a) a long ridge along the bed of a strait that diverts the flow of undersea currents, (b) an accumulated mass of sediment and rocks that has flowed out into the ocean from the mouth of a large river, or (c) a submerged, isolated mountain, usually of volcanic origin, rising above the ocean floor?

**23** Arches National Park in Utah, overlooking the gorge of the Colorado River, contains a vast array of formations that look as though some gigantic sculptor had been at work. These are in the shape of arches, windows, spires, and tall pinnacles. How were they formed?

**24** Visitors to Petrified National Forest in Arizona see trees that have been preserved in a unique fashion for millions of years. What does the term "petrified" mean?

**25** Scientists who study fossil plants and animals and their remains in various earth structures are known as: (a) botanists, (b) paleontologists, or (c) embryologists?

**26** Like many other canyons, the Grand Canyon was formed by the cutting action of a river (in this case the Colorado), erosion, and the upheaving of subterranean rocks over a long period of time. How long did it take for the mile-deep Grand Canyon to form: (a) one million years, (b) two million years, (c) 20 million years, or (d) more than 100 million years?

**27** The highest natural bridge in the world is near K'ashish, Sinkiang, China, rising more than 900 feet with a span of some 150 feet. For this two-parter: (a) what is a natural bridge, and (b) where is the longest natural bridge?

**28** Bryce Canyon National Park in the western United States is noted for its pink limestone cliffs which have been eroded over the years so they have taken on the shapes of miniature cities, cathedrals, and towers. In what state is Bryce Canyon?

**29** The mountain summit that is farthest from the earth's core is not Everest, but Chimborazo, which is only 20,560 feet above sea level but has an equatorial position that makes this record possible. In what mountain range is Chimborazo?

**30** Broad movements of the earth's crust are identified by the term: (a) epeirogeny, (b) geochronology, (c) hydration, or (d) sedimentation?

**31** The world's largest pothole is near Rockwood, Ontario, Canada, which also has the world's greatest concentration of potholes. Geologically, what is a pothole?

**32** Stalagmites and stalactites are distinctive features in caves. One hangs like an icicle from cave ceilings and the other projects up from cave floors. Which is which?

**33** The exploration of caves has become popular in recent years. What is this activity called?

**34** Ball's Pyramid near Lord Howe Island in the Pacific is a formation that is more than 1,840 feet high, but with a base that is only about 650 feet in circumference. What is the geological term for this kind of formation?

**35** The lithosphere is mentioned often in geography and geology books as a vital factor in the creation and development of landforms of all kinds. Is the lithosphere: (a) the air or gas surrounding the planet earth, (b) the force that molded the earth into a spherical shape millions of years ago, (c) a hot, hollow region at the center of the earth, or (d) the outer, relatively rigid part of the earth's crust?

**36** How would you define a coulee: (a) a spirelike rock jutting upward from a cliff or rock formation, (b) a ravine, gorge, or stream bed formed by an intermittent flow of water, or (c) an isolated plateau, usually not more than a mile long, projecting upward on a grassy plain?

**❸❼** When traveling through mountainous regions, you are likely to see any number of folds in the surface in which rock layers slope outward from the crest of a mountain. What are they called?

**❸❽** What is the definition of an *aiguille*: (a) a long, flat tableland high in the mountains, (b) a natural tunnel that has been eroded in the side of a mountain, or (c) a spire of rock projecting upward on a cliff or mountain?

**❸❾** Yellowstone National Park is noted for its impressive obsidian cliffs. What is obsidian?

## Answers

**①** *(c) The earth is slightly larger than the sixth planet, Venus, which is 7,550 miles in diameter, and considerably larger than Mars, which is 4,220 miles in diameter.*

**②** *An alluvial plain.*

**③** *(a) Badlands National Monument. (b) Badlands are striated (layered) gullies and hills and contain extensive fossils that help scientists determine the type of land-forming activities that took place in prehistoric times.*

**④** *The Gulf Stream originates in the Gulf of Mexico, swings around Florida,* and flows northeast along the East Coast. Off Cape Hatteras, North Carolina, it veers eastward and mixes with the North Atlantic Drift, which heads toward Europe.

⑤ Various forms of granite, some of which are the oldest known rocks on earth. Granite underlies practically all of the continent of North America.

⑥ Lake Bonneville, in northwestern Utah, which has six terraces that reveal the different water levels of the past and endless stretches of hard-packed lake bed.

⑦ A continental shelf.

⑧ Coral atolls, like Kwajalein in the Marshall Islands in the Central Pacific, were vital during World War II as air strips and military bases on the route to Japan.

⑨ The sea bed under the North Sea contains enormous deposits of oil and gas.

⑩ (a) Hudson Bay is both a bay and an inland sea. (b) It is located in the northeastern region of Canada, occupying an area of some 475,000 square miles, or about four-fifths the area of Alaska.

⑪ Continental Drift, the theory that the positions of the continents on the earth's surface have changed immensely through geo-

*logic time, with land masses splitting and moving apart as the seas were formed between them.*

⑫ *Sugarloaf Mountain is in Rio de Janeiro, Brazil.*

⑬ *Australia.*

⑭ *Near volcanoes, active or extinct. These rocks are formed by lava and filled, like sponges, with air pockets.*

⑮ *Wind Cave.*

⑯ *They are geysers. There are so many in Iceland that the residents of some towns seldom, if ever, have to use hot water heaters in their homes.*

⑰ *The Algarve.*

⑱ *You would see a whole forest of oil rigs sticking up from its waters. The geologists naturally spend time there exploring for more oil and gas.*

⑲ *Because the waters are too salty to support any form of life.*

⑳ *(a) An aquifer.*

㉑ *The Ice Age. The glaciers of this epoch carved many of our lakes and molded the shapes of mountains.*

㉒ *(c) A seamount is an undersea peak that more or less stands by itself.*

㉓ *By the action of water, frost, and wind over millions of years.*

㉔ *It means "stone-like." The tree trunks in this forest were literally transformed into stone through the action of minerals in the water into which they had fallen.*

㉕ *(b) Paleontologists.*

㉖ *(b) It has been estimated that the river and other forces of nature cut the level of the canyon about one foot every 384 years. At that rate, it would have taken two million years to mold the Grand Canyon into its present state.*

㉗ *(a) A natural bridge is a sandstone arch that has been formed by weathering and erosion. (b) The longest such bridge is near Moab, Utah.*

㉘ *Utah.*

㉙ *Chimbarazo is in the Andes of South America.*

㉚ *(a) Epeirogeny.*

㉛ *A pothole is a gouge or basin in bedrock that has been formed by the action of stones being swirled around in a river or by the movement of a glacier.*

㉜ *Stalactites are formed downward, like icicles, while stalagmites gradually rise upward, formed by deposits of minerals (usually calcium) dripping on them.*

㉝ *Spelunking. A "spelunker" is one who explores caves as a hobby.*

㉞ *It is a rock pinnacle.*

㉟ *(d) The lithosphere is the crust that holds not only the landforms but the oceans and seas.*

㊱ *(b) Coulee is derived from the French word* couler, *"to flow."*

㊲ *They are called anticlines.*

㊳ *(c) The term* aiquille *derives from a French word meaning "needle."*

㊴ *Obsidian is a volcanic glass, similar in structure to granite, but usually dark and translucent.*

# The Oceans
and Seas

Т*he accumulated oceans and seas on our planet cover more than seventy percent of the earth's surface. Water exists in as many forms as the somewhat more obvious landforms: bays and inlets, straits and channels, firths and gulfs, bights and loughs, estuaries and washes, and much, much more.*

*Let's get our feet wet and see how much you really know about the oceans and seas. We'll start the questioning the easy way, with some True/False questions that give you a fifty-fifty chance of being right, even if you're not sure.*

❶ True or false. The largest ocean is the Pacific.

❷ True or false. All the world's lands above sea level could fit into a space the size of the Pacific Ocean.

❸ True or false. The Indian Ocean is the second largest ocean, beating the Atlantic by a narrow margin.

❹ True or false. The average depth of the world's oceans is more than 11,000 feet.

❺ True or false. The greatest depth ever discovered in the oceans is just under 24,000 feet.

❻ True or false. The deepest parts of the ocean are commonly known as continental shelves.

❼ True or false. Sunlight penetrates to about 2,000 feet in the oceans, wherever the water is clear.

❽ True or false. In addition to the Gulf Stream, there are at least ten other major currents in the ocean.

❾ True or false. Ocean water is salty because of millions of fissures in the sea bottom that exude salty minerals.

❿ True or false. The Arctic Ocean is largely a solid mass of ice from sea bed to surface at and near the North Pole.

**11** As the name implies, three major currents that flow in east-west and west-east directions near the equator are the North Equatorial Current, the South Equatorial Current, and the Equatorial Counter Current. In which oceans would one or more of these be found?

**12** What do the following have in common: Juan de Fuca, the Bosporus, Mackinac, Skagerrak, Bass, Cook, and Dardenelles?

**13** What is the definition of the geographic feature these places have in common?

**14** If you were to sail the ocean waters around Scotland, you might pass the following places: Solway, Moray, Pentland, Lorn, and Clyde. What do they have in common?

**15** What is the meaning of the geographical term associated with the places in the preceding question?

**16** Ocean currents bend to the right when flowing north of the equator, while those flowing south of the equator bend to the left. What is the reason for this?

**17** In the oceans, spring tides are the highest tides. They occur when the sun's gravitational pull is combined with the moon's and thus forces the water higher. The lowest tides occur when the pull of the moon is at right angles to the sun and thus does not exert as much influence. What are these lowest tides called?

**18** Oceans are the home of the largest animals that have ever lived on earth, much larger than the dinosaurs. What is the largest animal of all?

**19** Some cities that were built along the sea were planned as ports and relied heavily on commercial shipping for their existence. One such city is situated in a lagoon that has suffered periodically from flooding. Its land area is sinking at the rate of about one inch each year, causing more and more flooding through the city's extensive canals. What is this ancient city? For extra credit, on what sea is it located?

**20** Amsterdam is another seacoast city which, for generations, has faced problems of flooding from the very sea that has been a major factor in its commercial existence. Where is Amsterdam located? What sea causes its chronic watery headaches?

**21** These steep-sided inlets of the sea that are characteristic of glaciated regions are found in Norway, Alaska, British Columbia, Scotland, Greenland, southern Chile, and southern New Zealand. What are they called?

**22** The beaches along the Atlantic coast of the United States are enjoyed for the sand and surf. The average swell on a typical summer day has a wavelength of 250 to 350 feet. How fast do the waves on the beaches along the Atlantic coast move shoreward: (a) 8 miles an hour, (b) 15 miles an hour, or (c) more than 20 miles an hour?

**23** The beginnings of marine studies can be traced back to the renowned Greek philosopher Aristotle, who made many accurate observations about the sea with no books or previous records to guide him. It was all the more remarkable because his findings occurred so long ago. In what century did Aristotle live and work?

**24** Biscayne Bay is a shallow, narrow inlet of the Atlantic Ocean in southeastern Florida. It is one of the major sites of vast tidal swamps of a type that are important to the environment because they trap sediments and, over periods of many years, serve to build up valuable land. What is the name of this type of swamp? (Hint: It is also the name of a tropical evergreen tree that grows in its waters.)

**25** One of the most famous bodies of water in the world is an arm of the Atlantic Ocean that separates France and England and has played a great part in the military and socioeconomic histories of both countries. What is this body of water called?

**26** Name the cities at each side of the narrowest point of the body of water mentioned in the preceding question.

**27** Where is the Bay of Cadiz, whose long history of seagoing activities dates back to the time of the Phoenicians, circa 1100 B.C.?

**28** One of the most famous bays in America, scenically, commercially, and historically, is the Chesapeake, explored and charted in 1608 by the English colonist, John Smith. Name at least one of the four major rivers that flow into Chesapeake Bay.

**29** What is the name for the lower part of rivers, streams, or drainage channels that are affected by the rise and fall of tides from an adjacent sea?

**30** The Bermuda Triangle has been the subject of several books and numerous articles during the past twenty-five years or so. Where in the Atlantic Ocean is it located, and why has it received so much publicity?

**31** In popular sea tales, what was the Sargasso Sea, with its relatively still waters, noted for that made mariners avoid it?

**32** People who like to swim in the ocean are often warned to avoid certain beaches because of the danger of strong, narrow streams made up of the fast seaward flow of water that has been swept shoreward by waves. What are these watery hazards?

**33** When we think of beaches, we usually picture sand that is white or buff-colored. But there are numerous beaches around the world that are black. What causes this?

**34** This sea was named after a king of Athens who, according to legend, threw himself into its waters when he erroneously thought that his son, Theseus, had been killed on a dangerous mission. What is this sea? (Hint: It separates Greece from Turkey.)

**35** In 1819 the British purchased an island on the southern tip of the Malay Peninsula which was named Singhapura and which they renamed Singapore. One of the greatest ports in Asia, this city lies on what sea?

**36** The North Sea is cold and bleak, yet of great economic importance to the nine countries that border its shores. Name these countries.

**37** What prominent body of water separates the Dominican Republic from Venezuela to the south?

**38** This body of water, which separates Maine and Nova Scotia, is known for its enormous tides, which can range as much as fifty feet between high and low water. What is its name?

**39** What two bodies of water does the Suez Canal connect?

**40** In northern waters, icebergs are constantly charted as soon as they are sighted to minimize the danger of collisions with ships. What is the meaning of the phrase "the tip of the iceberg?"

**41** What is the largest gulf in the world's oceans?

**42** What is the name of the formation that is recognized as the deepest part of the ocean?

**43** How many oceans are there? For extra credit, name them.

**44** What is the largest sea in the world?

**45** What is the smallest sea? Where is it located?

**46** Where is the Andaman Sea located?

**47** Three seas are named after colors. Name them, rank them in size, and give their locations.

**48** A well-known figure in the *Arabian Nights* was Sinbad the Sailor of Baghdad, who made seven spectacular voyages. On the first one he landed on what he thought was a small island only to discover it was a sleeping whale. This event supposedly took place while he was sailing the ocean lying closest to his country, Persia (now Iran). What ocean was that?

## Answers

① *True. The Pacific is by far the largest, both in surface extent and volume, covering 64,186,300 square miles.*

② *True. The Pacific would easily accommodate all of the land that can be seen on the surface of the earth.*

③ *False. The Indian Ocean is the third largest and is about twenty percent smaller than the Atlantic.*

④ *True. The average depth is 11,660 feet.*

⑤ *False. The greatest depth is just over 36,000 feet.*

⑥ *False. The deepest parts of the seas are oceanic trenches, which lie along the edges of plates (sections of the earth's crust) where one plate has been pushed down beneath another.*

⑦ *False. Even in the clearest water, at noon on a bright day, sunlight can penetrate no more than 1,000 feet.*

⑧ *True. All of the oceans have two or more important undersea currents providing continual movement of the water in well-defined directions and courses.*

⑨ *False. Ocean water is salty because rivers dissolve salts from the land and flush them into the sea. When seawater evaporates, the salts do not. They remain in the water and accumulate more and more, year after year.*

⑩ *False. The Arctic Ocean is largely covered by a vast ice pack. But it does not extend more than twenty or thirty feet below the surface and becomes even thinner in summer when the water temperatures rise.*

⑪ *They would be found in the Atlantic, the Pacific, or the Indian oceans, all of which have areas along the equator.*

⑫ *They are all straits.*

⑬ *A strait is a narrow body of water connecting two large bodies of water.*

⑭ *They are all firths.*

⑮ *A firth is the Scottish term applied to an arm of the sea, usually an estuary or strait.*

⑯ *This happens because the earth spins from west to east and that affects the course of ocean currents.*

⑰ *Neap tides. When a ship is "neaped," it is bottled up in a shallow harbor out of which it cannot sail until the next spring (high) tides come along.*

⑱ *The blue whale, also known as the sulfur-bottom, which can reach more than 100 feet in length and weigh more than 120 tons.*

⑲ *The city is Venice, which has been a major port for more than 1,500 years. It is in northeastern Italy at the upper reaches of the Adriatic Sea.*

⑳ *Amsterdam is the largest city of the Netherlands and is situated on the North Sea.*

㉑ *Fjords. They differ from most inlets and estuaries in that they are bordered by high, sheer walls that often extend far below the surface of the water.*

㉒ *(c) Average swells move from the open ocean shoreward at a speed of twenty to twenty-five miles per hour.*

㉓ *Aristotle lived in the third century, B.C.*

㉔ *Mangrove swamp,* home of the mangrove tree. *This type of ocean swamp is said to have added almost 2,000 acres of new land to the Florida coastline alone over the past fifty years.*

㉕ *The English Channel.*

㉖ *Dover, England, and Calais, France, are the two cities.*

㉗ *The Bay of Cadiz is located in the southwestern part of Spain in the Andalusian region.*

㉘ *The rivers flowing into the Chesapeake are the Susquehanna, Potomac, Rappahannock, and James.*

㉙ *An estuary.*

㉚ *The Bermuda Triangle is bounded by Bermuda, Puerto Rico, and a point near Melbourne, Florida. This triangle has become infamous in the annals of geography as an area that has had far more than its share of aircraft disappearances and ship losses, often under mysterious circumstances.*

㉛ *The Sargasso Sea was supposedly a place where huge masses of seaweed would entangle ships and prevent them from escaping. Although the tales were greatly exaggerated, the Sargasso is biologically recognized for its abundance of gulfweed and the rich marine life living within it.*

③② *Rip currents (riptides).*
*These forces can sweep unsuspecting victims off their*
*feet and carry them dangerously beyond their depth.*

③③ *These beaches, such as*
*one south of San Jose in Costa Rica, derive their*
*color from black rocks of volcanic origin that, over*
*thousands of years, have been ground into fine par-*
*ticles and washed along the shore.*

③④ *The Aegean Sea.*

③⑤ *The South China Sea*
*near the Strait of Malacca.*

③⑥ *England, Scotland,*
*Norway, Sweden, Denmark, Germany, the Nether-*
*lands, Belgium, and France.*

③⑦ *The Caribbean Sea.*

③⑧ *The Bay of Fundy.*

③⑨ *It connects the Mediter-*
*ranean with the Gulf of Suez and then the Red Sea.*

④⓪ *About nine-tenths of an*
*iceberg lies hidden below the sea. Thus, an iceberg*
*that had a visible mass above water of four cubic*
*miles would be about forty cubic miles in actual size.*

④① *The Gulf of Mexico. It*
*is more than 1,100 miles long and has an area of*
*about 700,000 square miles (almost as large as the*
*Caribbean Sea).*

(42) *The Marianas Trench in the Pacific Ocean, southwest of Guam. It is 36,198 feet at its deepest point.*

(43) *Five: Antarctic, Arctic, Atlantic, Indian, and Pacific. Some geographies list seven, referring to the North Pacific, South Pacific, North Atlantic, and South Atlantic.*

(44) *The Mediterranean. It covers 1,145,000 square miles.*

(45) *The Baltic Sea, which covers 160,000 square miles. It is located in northern Europe between the Scandinavian countries and Russia.*

(46) *In the Indian Ocean, off the southern coast of Asia.*

(47) *The largest, the Yellow Sea (480,000 square miles), is located off the east coast of Asia. The Red Sea (170,000 square miles) is located between North Africa and the Arabian Peninsula. The Black Sea (164,000 square miles) is located between southeastern Europe and southwestern Asia.*

(48) *The Indian Ocean.*

# Mountains, Valleys, and Other Landforms

Y*ou can rise to the occasion now by identifying some mountains, their locations, and interesting facts about them.*

*If you were to circle the earth in a spacecraft, you would be astonished at how many mountain ranges and peaks there are on our planet, and how much they contribute to the geographical profile of the world we live in. And bear in mind that, even from this superb vantage point, there would be many huge mountains you would not see at all because they lie hidden beneath the seas and oceans.*

*This chapter will also descend into the valleys around the mountains, as well as exploring plateaus, steppes, plains, promontories, cliffs, and other common landforms.*

*Let's start at the upper level, with mountains.*

❶ What is the highest point on earth: (a) Mount McKinley (Alaska), (b) Mount Kilimanjaro (Tanzania), or (c) Mount Everest (Tibet/Nepal)?

❷ How many mountain peaks are there in the world that rise more than 24,000 feet: (a) 22, (b) 58, or (c) more than 100?

❸ The greatest mountain system is known as the Mid-Atlantic Ridge, but most of its peaks have never been climbed. Why is that?

❹ The tallest active volcano is named Cotopaxi. Where is it located?

❺ What is the Appalachian Trail and where is it located?

❻ The Blue Ridge Mountains are ideal for hiking. If you were to walk their entire length: (a) how far would you hike, and (b) what states would you pass through?

❼ If you were to spend your vacation in the mountains mapping a *cordil-*

*lera*, you would be traveling: (a) across three mountains grouped together, (b) along an entire range of mountains, or (c) around and up a very high, snow-capped peak?

**8** Let's try a Matching Question. Match the highest points in the list on the left with the states in which they are located:

| **Mountain** | **State** |
|---|---|
| **1.** Mount Whitney | **(a)** Alaska |
| **2.** Mount McKinley | **(b)** Oregon |
| **3.** Mount Hood | **(c)** California |
| **4.** Mount Washington | **(d)** New York |
| **5.** Mount Marcy | **(e)** New Hampshire |

**9** Now match the mountains in the list on the left with the countries in which they are located:

| **Mountain** | **Country** |
|---|---|
| **1.** Mont Blanc | **(a)** Argentina |
| **2.** Mount Cook | **(b)** India |
| **3.** Matterhorn | **(c)** France |
| **4.** Aconcagua | **(d)** New Zealand |
| **5.** Nanda Devi | **(e)** Switzerland |

**10** Three well-known mountains have romantic names: Kilimanjaro, "Mountain of the Demon of Cold;" Ruwenzori, "Cloud King;" and Kenya, "Mountain of the Mist." They are all very high, very rugged, and located on the same continent. Can you name the continent and their general location?

**⑪** The Rocky Mountains extend from where to where?

**⑫** The White Mountains in New Hampshire are distinctive in that they are named for early American presidents. Which is the highest of the White Mountains?

**⑬** The Black Hills received their name from heavily forested slopes that, from a distance, appear to be black. Where are the Black Hills?

**⑭** Can you name any mountain at the North Pole?

**⑮** Try to match these mountain ranges with the continents on which they are found.

| Mountain | Continent |
|---|---|
| **1.** Carpathians | **(a)** Antarctica |
| **2.** Andes | **(b)** Asia |
| **3.** Kunlun | **(c)** North America |
| **4.** Sierra Nevada | **(d)** Europe |
| **5.** Elsworth | **(e)** South America |

**⑯** Mount Ararat is mentioned in the Bible and is thought to have been the place where Noah's Ark came to rest. Where is Mount Ararat?

**⑰** In Greek mythology, Erebus was the personification of darkness. Why was this an appropriate name for Mount Erebus, discovered by Sir James Ross in the 1840s when he was exploring the Antarctic?

**18** What do these mountains have in common: Cameroon, Nyirangongo, Karthala, and Piton de la Fournaise?

**19** The most famous mountain in Japan is visited by millions of Japanese regularly and is thought to be sacred. What is its name?

**20** What is the longest continuous mountain range in the world?

**21** What is the greatest high plateau in the world?

**22** Where are the following high mountains located: Weisshom, Taschhorn, Nadelhorn, and Lenzpitze?

**23** Australia is generally very low and flat and not known for its mountains as much as for other geographical and geological features. However, a popular Australian tourist attraction is an enormous dome mountain made of solid rock located in the middle of an almost endless plain. What is the name of this strange feature?

**24** Travelers to Peru who are anywhere near the town of Cuzco are more than likely to take a tour up into the Andes to view a mountain citadel that was built by the ancient

Incas and is one of the most remarkable archaeological ruins in the Americas. What is the name of
this famous place?

**25** Most people are famil-
iar with the Swiss Alps, but there are two other
mountain systems in the world that are referred to
as the Alps, and they are far from Switzerland.
Where are they?

**26** The Matterhorn is per-
haps the most familiar mountain in Europe. Where
exactly is it located?

**27** Four large American
cities are located in mountain regions at altitudes
of more than 6,000 feet. Can you name at least two
of them?

**28** Another major city
that is high in the mountains (though not quite so
high as the four mentioned above) is Denver,
Colorado. It's nickname indicates just how high it
is. Do you know it?

**29** One North American
capital is even higher in the mountains than any
city in the United States. Do you know the country
and the capital?

**30** Mount Washington, in
the White Mountains of New Hampshire, is not
particularly high by world standards, only 6,288
feet. However, it has one of the world's most

unique modes of transportation. Can you describe it?

**31** "Shangri-La" was the name of a magnificent mountain land used as the background for James Hilton's popular novel, *Lost Horizon*. Is this place real or fictional?

**32** Mount Vernon, just south of the nation's capital, is widely known as the historic home of President George Washington. Is there a mountain in the vicinity with that name?

**33** The most impressive cliff face in North America is said to be the northwest face of Half Dome, which stretches for 3,200 feet and is almost 2,300 feet high. In which one of our large national parks is Half Dome located?

**34** What is the word used for a cliff, headland, or hill with a broad, steep face?

**35** The Dolomites, or Dolomite Alps, in northern Italy are popular with mountain climbers because they are high enough (ranging up to almost 11,000 feet), but very reliable for climbing. One reason for this is found in the name itself. What is *dolomite*?

**36** If you drove up a high mountain to photograph an especially scenic *caldera*, you would expect to see which of the following: (a) a dramatic-looking cliff face, (b) a large crater, perhaps filled with water, or (c) a narrow pass through the mountain?

**37** Here's an easy one: Mount Rushmore in the Black Hills of South Dakota is neither unusually high nor ideal for climbing or camping. Yet it attracts thousands of visitors each year because of the busts of four American presidents carved into its face. Can you name them?

**38** The Pan American Highway is not in any sense a "superhighway" as we know it, but rather a linkage of many roads ranging from four-lane highways to narrow dirt roads, many of them in the mountains. Along what mountain range does most of this highway run?

**39** The Catskill Mountains were made famous by author Washington Irving. Where are the Catskills?

Now let's descend from the mountains and explore some other common landforms.

**40** What is the name for an area of land almost completely surrounded by water except for a narrow strip?

**41** Valleys are most commonly associated with agriculture because of the presence of streams, richer soil, and walls that afford protection from the elements. In California there is a valley noted for its grapes that produce wine. Name the valley.

**42** If you were to fly in a small plane over the Soviet Union and southeastern Europe, you would see many examples of a typical land formation called a *steppe*. What is a steppe: (a) temperate grasslands consisting of level, generally treeless plains, (b) a series of plateaus rising up to a mountain, or (c) very narrow valleys, some with streams?

**43** The landforms of the world, high and low alike, are covered with forests of many sizes and species. Generally, however, they fall into just two types, evergreen and deciduous. (a) Define evergreen trees, and (b) define deciduous trees.

**44** You have probably noticed that there is a level along high mountains beyond which trees will not grow and the foliage is limited to low, scraggly shrubs, grasses, and often certain kinds of wild flowers. What is the name for this particular elevation: (a) the lithosphere, (b) the highlands, or (c) the timberline?

**45** The surface of the land in some regions is marked by extensive moraines. Is a moraine: (a) a large pothole that resembles something on the moon's surface, (b) an outcropping of rock that has been thrust up from below the surface and that bears evidence of volcanic action in prehistoric times, or (c) rock and soil debris carried and finally deposited by glaciers after they have melted, often in the form of long ridges?

**46** What do the following have in common: Good Hope, May, Kennedy, York, and Verde?

**47** What is the meaning of the geographical term that is the answer to the preceding question?

**48** One of the most famous valleys in Europe is the Loire. For a two-parter: (a) where is it, and (b) what is it noted for?

**49** What do the following have in common: Panama (the country), Suez (in northeastern Egypt), and Corinth (in southeastern Greece)?

**50** What is the meaning of the term used to describe the landforms in the preceding question?

**51** What do the following have in common: Daryal (in the Caucasus), Khyber (on the Pakistan-Afghanistan border), and Simplon (in Switzerland)?

**52** How would you describe the landform that characterizes the places in the preceding question?

**53** Extensive stretches of land, generally unsuited to crop cultivation but providing native grasses and other forage plants for livestock grazing, are found in many places around the world. They are especially characteristic of parts of the western United States and Canada. Are these lands: (a) broadlands, (b) ranges, or (c) scarps?

**54** One of the highest fertile valleys in the world lies at 4,500 feet above sea level surrounding the ancient city of Katmandu, on the old trade route between Tibet and India. In what small kingdom is this located?

**55** Any region where water drains down the slopes to a river, lake, or other body of water is called: (a) a tributary, (b) an extrusion, or (c) a watershed?

**56** Rhodesia and the Republic of South Africa are noted for their grassy, undulating plateaus that are fertile enough to support numerous kinds of crops and livestock. What are these plateaus called?

**57** What do we call the line of contact between land and water surfaces?

**58** What word indicates a strip of land of indefinite width that is landward of the shore?

**59** The Connecticut River Valley was an area that attracted early settlers in America because of the rich land along its shores and the availability of water transportation. What four states share portions of this fertile valley, which stretches for some 400 miles?

**60** The Palisades of New Jersey along the western shore of the lower Hudson River are graphic examples of long, steep cliffs or bluffs caused by the erosion of tilted rock layers. What is their proper term?

**61** In some romantic English novels of the nineteenth century, the settings were often described as being on, or near, a type of open, peat-rich wasteland. What is this desolate tract of land called?

**62** What two Asian countries share the heavily populated Indus valley, whose major crops are wheat, rice, dates, corn, millet, and fruits?

**63** The Valtellina is an Alpine valley of the upper Adda River in Lombardy, northern Italy, extending from Lake Como to the Stelvio Pass. The valley is a very picturesque and fertile agricultural region. For what major crop is it chiefly noted?

**64** The Valley of the Kings and the Valley of the Queens are noted more for their archaeological wonders than for their soils and crops. In what country are these valleys located? For extra credit, what is their archaeological importance?

**65** Match these fertile river valleys with the countries that most depend upon their crops:

| Valley | Country |
|--------|---------|
| **1.** Irawaddy | **(a)** New Zealand |
| **2.** Murray | **(b)** Ireland |
| **3.** São Francisco | **(c)** Burma |
| **4.** Rakaia | **(d)** Australia |
| **5.** Shannon | **(e)** Brazil |

## Answers

**①** *(c) Mount Everest, in the Himalayas, is the highest spot on earth, at 29,028 feet.*

**②** *(c) The world has 109 peaks that rise above 24,000 feet. All of them are in Asia and 96 of them are in the Himalaya-Karakoram Range.*

**③** *Because most of the peaks of the Mid-Atlantic Ridge lie beneath the ocean.*

**④** *Cotopaxi is in Ecuador, in the Andes of South America.*

⑤ *The Appalachian Trail is the world's longest continual hiking path, stretching along the ridges of mountains in fourteen states in the eastern United States, from Mount Katahdin in Maine to Springer Mountain in Georgia. It is America's only official National Scenic Trail.*

⑥ *(a) You would hike well over 500 miles, their length as the crow flies. (b) You would pass through West Virginia, Virginia, North Carolina, eastern Tennessee, and the northern tip of Georgia.*

⑦ *(b) A* cordillera *("range of mountains" in Spanish) is the name applied to any extensive group or system of mountains.*

⑧ *1(c) Mount Whitney/ California, 2(a) Mount McKinley/Alaska, 3(b) Mount Hood/Oregon, 4(e) Mount Washington/New Hampshire, and 5(d) Mount Marcy/New York.*

⑨ *1(c) Mont Blanc/ France, 2(d) Mount Cook/New Zealand, 3(e) Matterhorn/Switzerland, 4(a) Aconcagua/Argentina, and 5(b) Nanda Devi/India.*

⑩ *They are all in Africa, located right along the equator.*

⑪ *The Rocky Mountains run from Alaska southwest to central New Mexico.*

⑫ *The highest is Mount Washington at 6,293 feet.*

(13) *The Black Hills are in South Dakota and Wyoming.*

(14) *This is a trick question. There are no mountains at the North Pole, or any kind of landform either. The entire north polar region is covered with the waters of the Arctic Ocean.*

(15) *1(d) Carpathians/Europe, 2(e) Andes/South America, 3(b) Kunlun/Asia, 4(c) Sierra Nevada/North America, and 5(a) Elsworth/Antarctica.*

(16) *Actually, there are two such mountains, Little Ararat, almost 13,000 feet high, and Great Ararat, whose peak stands at almost 17,000 feet. They are in eastern Turkey, near the Soviet and Iranian borders.*

(17) *It was appropriate because the Antarctic lies in total darkness for six months of the year.*

(18) *They are all active volcanoes in Africa and have erupted between 1977 and 1991.*

(19) *Mount Fujiyama, also known as Mount Fuji.*

(20) *The Andes, which extend more than 5,000 miles down the west coast of South America.*

㉑ *The Tibetan Plateau in central Asia, which has an average altitude of 16,000 feet.*

㉒ *They are all mountains in Switzerland.*

㉓ *Ayers Rock, near the town of Alice Springs in the Northern Territory of Australia. It rises 2,820 feet above the flat, arid countryside, resembling an enormous oval bowl turned upside down. Some refer to it as a "beached whale."*

㉔ *Machu Picchu (pronounced* ma-choo-pek-choo)*, which is one of the most dramatic sites in South America and looks down on the Urubamba River some 2,000 feet below.*

㉕ *One is the Australian Alps, a chain running along the southeastern coast of Australia. The other is the Southern Alps, located on South Island in New Zealand, which extend almost the entire length of the island and rise to heights of more than 12,000 feet, culminating with snow-capped Mount Cook at 12,349 feet.*

㉖ *It is near Zermatt on the Italian border of Switzerland.*

㉗ *Cheyenne, Wyoming, 6,100 feet; Flagstaff, Arizona, 6,900 feet; Gallup, New Mexico, 6,540 feet; and Santa Fe, New Mexico, 6,950 feet.*

㉘ *"The Mile High City," at exactly 5,280 feet.*

㉙ *Mexico City, the capital of Mexico, which is situated at 7,347 feet.*

㉚ *Mount Washington has a cog railway, with a steam locomotive that grinds its way up and down the tracks through use of a cogged center driving wheel that meshes with a similarly cogged third rail.*

㉛ *"Shangri-La" is fictional, but was supposedly based on the author's knowledge of the mountains of Tibet. Tibet is situated on a high plateau averaging some 16,000 feet.*

㉜ *No. The estate was named in honor of Admiral Vernon of the British navy, under whom George Washington's half brother served. The "Mount" simply refers to the hilly slope on which the main house sits.*

㉝ *Yosemite, in California.*

㉞ *A bluff. This formation is found along the ocean shores, but also on inland plains, lakes, and rivers.*

㉟ *Dolomite is a carbonate rock that is similar to limestone, but harder and heavier. It is compatible with mountain-climbing equipment and will not easily break or shatter.*

㊱ *(b) A* caldera *is a crater, often formed by volcanic action if on a mountain top or by the impact of meteorites, if found on plateaus or lowlands.*

(37) *Carved on the face of Mount Rushmore, and visible for sixty miles, are the busts of George Washington, Abraham Lincoln, Thomas Jefferson, and Theodore Roosevelt. It took fourteen years to carve the figures, which were completed in 1941.*

(38) *The Pan American Highway runs along the Andes of South America.*

(39) *The Catskills lie along the Hudson River in southeastern New York State.*

(40) *A peninsula.*

(41) *The Napa Valley.*

(42) *(a) A steppe is similar to a prairie.*

(43) *(a) Evergreens are trees that retain their needles or scale-like leaves throughout the entire year, the shoots of the past season not being shed until new growth has been formed. (b) Deciduous trees are trees whose leaves fall annually during one season of the year and are not replaced until a following season.*

(44) *(c) The timberline. Its elevation varies according to the latitude, the climate, exposure to sunlight, prevailing winds, and other factors.*

(45) *(c) From the Spanish, morena, a pile of debris. These are sometimes referred to also as drumlins.*

㊻ *They are all capes.*

㊼ *A cape refers to a point of land jutting out into the sea.*

㊽ *(a) The Loire Valley is in France. (b) It is also called the "Chateau Country" because of the many historical chateaux built here. The Loire Basin is especially rich in gardens and vineyards.*

㊾ *They are all isthmuses.*

㊿ *An isthmus is a narrow neck of land connecting two larger land areas.*

(51) *They are all passes through the mountains.*

(52) *A pass is an opening or way by which a natural barrier, such as a mountain range, can be traversed.*

(53) *(b) These extensive stretches of land are called ranges.*

(54) *Nepal.*

(55) *(c) It is called a watershed and is found on a mountain slope or other elevated region.*

(56) *These fertile plateaus are called* velds *or* veldts.

(57) *The line of contact between land and water surfaces is called the shoreline.*

**⑤⑧** *The coast.*

**⑤⑨** *Connecticut, Massachusetts, New Hampshire, and Vermont.*

**⑥⓪** *Escarpments, sometimes called scarps.*

**⑥①** *A moor.*

**⑥②** *India and Pakistan.*

**⑥③** *Grapes, for the production of fine Italian wines.*

**⑥④** *These valleys are located in Egypt. They are the sites of the tombs of ancient rulers and the source of priceless historical relics.*

**⑥⑤** *1(c) Irawaddy/Burma, 2(d) Murray/Australia, 3(e) São Francisco/Brazil, 4(a) Rakaia/New Zealand, and 5(b) Shannon/Ireland.*

# Lakes, Rivers, and Other Freshwater Bodies

A lake is a lake, a river is a river, a pond is a pond. Or are they? Bodies of freshwater, whether running or still, defy classification because they are so varied and regional. A "lake" in Iowa may be a "pond" in New England. A "river" in Ireland may be a "stream" in Brazil. In arid lands, a little water looks like a lot. In rain-drenched climates, large bodies of water are so commonplace as to be inconsequential in the eyes of the beholder.

So let's go fresh-water exploring and see what we can discover.

**❶** The scientific study of lakes and other bodies of fresh water is known as: (a) acquatology, (b) limnology, or (c) paleontology?

**❷** The Great Lakes form the largest body of fresh water in the world and, with their connecting waterways, are the largest inland water transportation entity. Which one of these lakes lies wholly within the United States, instead of being shared with Canada?

**❸** Man-made lakes, created by dams and technically known as reservoirs, are common around the world. Can you match the following such lakes (among the twenty-five largest on earth) with the countries in which they are located?

| Lake | Country |
|------|---------|
| **1.** Aswan | **(a)** Turkey |
| **2.** Akosombo | **(b)** Russia |
| **3.** Lake Mead | **(c)** Ghana |
| **4.** Ataturk | **(d)** Egypt |
| **5.** Bratsk | **(e)** United States |

**❹** Some lakes are created when a river forms a loop that eventually is cut off from the main stream and becomes a still body of water on its own. What is the name applied to this kind of lake?

**❺** What is the difference between a *lake*, a *lough*, a *loch*, and a *tarn*?

**❻** Loch Ness, which is twenty-two miles long and very deep for a medium-sized lake (700 feet), is most noted for the monster that supposedly inhabits its waters. Just where is Loch Ness?

**❼** The Lake District, surrounded in pastoral beauty by the gentle Cumbrian Mountains and verdant forests, is known for its fifteen picturesque lakes. Where is the Lake District?

**❽** The following lakes are among the twenty largest in the world. See if you can match the lake with the country in which it is located.

| **Lake** | **Country** |
|---|---|
| **1.** Victoria | **(a)** Australia |
| **2.** Eyre | **(b)** Venezuela |
| **3.** Great Slave | **(c)** Russia |
| **4.** Maracaibo | **(d)** Uganda-Tanzania-Kenya |
| **5.** Ladoga | **(e)** Canada |

**❾** Excluding the Great Lakes, thirty other lakes in the United States each cover an area of one-hundred square miles or more. Listed below are the five largest of these. Rank them by size in descending order: Great Salt Lake, Utah; Lake Champlain, New York-Vermont; Tahoe, California-Nevada; Okeechobee, Florida; and Pontchartrain, Louisiana.

**❿** If you were sailing through the five Great Lakes, which would you pass, in order, from east to west? For extra credit, which is the largest?

**⓫** Lac Leman is a crescent-shaped lake, ringed by mountains. Located in western Switzerland, on the French border, it is some forty-five miles long and has a maximum depth of more than one thousand feet. By what more common name is it known?

**⓬** Nearly fifty percent of the world's lakes, in terms of numbers, not size, are located in a single country. Which of the more than 160 countries in the world has this distinction?

**⓭** The largest underground lake yet discovered in North America, which covers almost five acres, lies some 300 feet underground near Sweetwater, Tennessee, in Craighead Caverns. Name the lake.

**⓮** The lake that is perhaps the most recognized of any in North America for its scenic beauty is Lake Louise. It is surrounded by high peaks, glaciers, and snow fields that are reflected in its clear waters. Where is Lake Louise?

**⓯** Mammoth Cave in southern Kentucky is one of the largest known caves in the world and contains an underground lake that is fed by numerous underground streams. Although there are fish in the lake, they are different from fish you might find in a surface lake. What is the difference?

**16** Walden Pond is a very tiny body of water that has become famous in books about New England and American history. Where is Walden Pond located?

**17** Crater Lake, the second deepest lake (1,932 feet) in North America, is unique in another way too: it has no inlets or outlets. Why doesn't it dry up or become totally stagnant and dead?

**18** Vacationers in the United States who enjoy lakes for boating, fishing, or scenery, generally go to the northern states rather than the southern ones. Apart from matters of climate or season, why is this so?

**19** Lake Placid attained renown as the site of the 1932 Winter Olympics, the first to be held in the United States. An internationally known resort and winter sports center, where is Lake Placid located?

**20** What state holds the record for the most fresh water (lakes, ponds, and other bodies) within its borders?

**21** The largest swamp in the world covers an area of some 18,000 square miles, about the size of Vermont and New Hampshire combined. In what country is it located: (a) the Philippines, (b) Canada, (c) Australia, or (d) the Soviet Union?

**22** One of the largest artificial lakes in the world, measured in terms of surface area, is Lake Volta, created by construction of the huge Akosombo dam. In what country is it located?

**23** Can you name two or more of the four countries that have shorelines on Lake Tanganyika, the second largest lake in Africa?

**24** What is a *mere*?

**25** Many American metropolises contain small bodies of water in the heart of the city. Try to match the following cities with the bodies of water you would find in them:

| City | Lake |
|---|---|
| **1.** New York | **(a)** White Rock Lake |
| **2.** Washington, D.C. | **(b)** Lake on the Commons |
| **3.** Boston | **(c)** Blue Lagoon Lake |
| **4.** Miami | **(d)** The Tidal Basin |
| **5.** Dallas | **(d)** Central Park Lake |

**26** Which large American city has the greatest number of natural·lakes within its environs?

Since the earliest days of adventure and exploration, rivers have been the keys to unlocking the secrets of unknown or little-known territories. With the development of trade, rivers became the arteries that made it possible to establish cities far from the ocean shores, and to transport produce and equipment with relative ease and minimum expense. Those nations fortunate enough to have numerous navigable rivers tended to develop more quickly than those that had to rely on extensive land transportation or to limit their commerce to coastal regions.

So let's switch the questions to test your knowledge of the world's rivers and other waterways, great and small.

**27** Paradoxically, the longest river in Italy has the shortest name, consisting of only two letters. It begins in the mountains near Switzerland, runs past Torino (Turin) and Cremona, and empties into the Adriatic Sea north of Ravenna. Can you name it?

**28** Another famous Italian river is the Tiber, which runs about 250 miles from the heights of the Apenine Mountains to its mouth. What sea does it empty into? For extra credit, what city is most associated with it in history, legend, and literature?

**29** When we think of this River, we picture Paris, the Eiffel Tower, and everything that is French. For a two-parter: (a) name the river, and (b) tell what body of water it flows into.

**30** We often hear about the "Left Bank" of the river in the preceding question, where artists exhibit their paintings and people stroll contentedly on warm spring days. In terms of the compass point, which side of the river is the *left* bank?

**31** What river flows through a city that has a country inside it?

**32** What river is called "The Mother of Rivers"?

**33** It is totally Russian and the longest river in Europe. Can you name it?

**34** Another very famous river starts as a tiny spring in the Cotswold hills in England, and was used by Julius Caesar's legionaries for the water supply for their nearby Roman fort. For a three parter: (a) name the river, (b) give the pronunciation, and (c) pinpoint the body of water into which it flows.

**35** Although Lisbon, the capital of Portugal, is seven miles inland from the ocean, it has been known for centuries as a major port. Can you name the river that has made this possible? Also, name the body of water into which it flows.

**36** The Continental Divide is the source of many rivers, such as the Colorado, which start as tiny streams and later become large, fast, and turbulent. Where is the Continental Divide on the map?

**37** The great Mississippi runs like a huge artery through the heartland of the United States from its origin in Lake Itasca, Minnesota, to its mouth near New Orleans in the Gulf of Mexico. Two other large rivers join the Mississippi, one from the east and one from the west. What are the names of these two rivers? For extra credit, where do they meet the Mississippi?

**38** One major river, more than 1,000 miles long, is almost synonymous with Alaska, although it starts in British Columbia before crossing the forty-ninth state from east to west and emptying into the Bering Sea. Can you name the river?

**39** It was discovered by the French explorer, Jacques Cartier, in 1535 when he sailed westward from the Atlantic looking for the Northwest Passage. Today, it is noted, among other things, for the multitude of islands in its waters. (a) What is the name of this river, and (b) through which two major Canadian cities does it flow?

**40** The Potomac is most recognized as the river that flows through Washington, D.C., and past George Washington's home at Mount Vernon. Can you name (a) the mountain range in which it originates, and (b) the body of water into which it flows at the end of its 290-mile course?

**41** Known as one of the most devious rivers in the United States, this river starts its 650-mile course in the Appalachian Mountains in Virginia and eventually joins the Ohio River just before that river joins the Mississippi. Name the river.

**42** The Mekong River has frequently been in the news as the chief river of the Indochina Peninsula. Originating in the Tibetan Plateau, it flows through five peninsular countries that rely on it for transportation, irrigation, and military operations. Can you name two or more of these five countries?

**43** What river forms more than half of the border between the United States and Mexico.

**44** Although the continent of Australia is almost as large as the United States in land area, it is unique in that it has only one important river system, which consists of two rivers and their relatively small tributaries. Can you name the two rivers?

**45** What is the longest river in the world? For extra credit, what is the runner-up?

**46** As might be expected, the massive mountains of Tibet, with their constant snows and storms, spawn streams that eventually turn into mighty rivers. Two of them flow through China and eventually come close together where they are joined by China's Grand Canal. (a) What are the two rivers, and (b) what large city lies at the southern end of this canal?

**47** What is the name of the river that originates in the Carpathian Mountains in southern Poland and flows northward through Krakow and Warsaw for 630 miles to the Baltic Sea?

**48** What river in the western United States is known as the "Big Muddy?"

**49** What major American river originates right in the heart of a large city that has a population of more than half a million people?

**50** Thirteen years before the Pilgrims landed at what is now Plymouth, Massachusetts, a group of pioneers led by Captain John Smith sailed up a river far to the south to found a settlement in what was later to be named Virginia. For a two-parter: (a) what is the name of this river, and (b) what is the name of the town that evolved from the original colony?

**51** If you were to travel to the heart of this famous African river that orginates near Lake Tanganyika, you might see a wide range of animal life, including hundreds of species of fish, water snakes, sharks, stingrays, crocodiles, and hippos. What is the name of this river that is, at almost 3,000 miles, the seventh longest in the world?

**52** The Red River, appropriately named because it is distinctively colored by red clay along its upper course, flows eastward, then south to join the Atchafalaya and the Mississippi rivers in southern Louisiana. Along its middle section, it forms the ragged border between two states. Can you name those states?

**53** What is the name of the town where Shakespeare lived? Also, name the river that flows through it.

**54** This river in New York State starts as a tiny trout stream on Mount Marcy, at 5,300 feet the highest mountain in the state. By the time it has reached its mouth at New York City, 300 miles to the south, it is extremely broad, mingling with the salt water of the Atlantic. What is the name of this river?

**55** Let's try some two-parters. Here's the first: (a) where is the Somme River, and (b) what war was most closely associated with it?

**56** (a) Where is the Yalu River? (b) During what war was it the location for strategic military actions?

**57** (a) What are the names of the two rivers that join to form the Nile? (b) At what place does this juncture occur?

**58** (a) What river is locally referred to as the *Rio Bravo del Norte*? (b) How long is it?

**59** Canals have been around for many hundreds of years, usually constructed to join rivers and/or other bodies of water. What is the longest canal system in the world?

**60** Most major canals have locks. What is a lock and why is it necessary?

**61** What famous canal was constructed with a man-made lake in the middle?

**62** Lost River is a popular tourist attraction in the White Mountains of New Hampshire. Why is it "lost"?

**63** A river in Africa is known by the natives as "The Strong Brown God." Can you name the river?

**64** What river continues far beyond its mouth, thus making it possible for boats to navigate in fresh, not salt, water as far as 150 miles out at sea?

**65** A major city is located on an island in the St. Lawrence River. Name it.

**66** The Jordan River was named after a Hebrew word meaning "to descend" because it drops so rapidly to a low-lying sea. What is that sea?

**67** When the Greeks began exploring lands near the western end of the Mediterranean Sea, they discovered a river, which they called the Iberus (now the Ebro), and referred to the region as Iberia. In geographical terms, what is Iberia? For extra credit, what countries lie within its borders?

**68** Match these other dominating rivers of the world (not previously mentioned) with the countries with which they are most closely associated:

| Lake | Country |
|------|---------|
| **1.** Lena | **(a)** Brazil |
| **2.** Orinoco | **(b)** Turkey |
| **3.** Madeira | **(c)** Canada |
| **4.** Mackenzie | **(d)** Soviet Union |
| **5.** Euphrates | **(e)** Venezuela |

## Answers

**1** *(b) Limnology, which derives from a Greek word meaning "pool" or "marsh."*

② *Lake Michigan is the only one of the five lakes that lies wholly within the United States.*

③ *1(d) Aswan/Egypt, 2(c) Akosombo/Ghana, 3(e) Lake Mead/United States, 4(a) Ataturk/Turkey, and 5(b) Bratsk/Russia.*

④ *An oxbow lake. The Mississippi, which twists and turns and contantly changes course, is noted for creating oxbows.*

⑤ *Very little. They are all lakes. The word* lough *is commonly used in Ireland and* loch *in Scotland. They are pronounced the same.*

⑥ *Loch Ness is in north central Scotland in a fault formation known as the Great Glen, formed by glaciers during the Ice Age. The deep rifts in the earth's surface, going hundreds of feet below sea level, support the controversial theory that strange creatures from ancient times could have survived in this environment.*

⑦ *The Lake District is in northwestern England.*

⑧ *1(d) Victoria/Uganda-Tanzania-Kenya, 2(a) Eyre/Australia, 3(e) Great Slave/Canada, 4(b) Maracaibo/Venezuela, and 5(c) Ladoga/Russia.*

⑨ *The largest of these*

*lakes is Great Salt Lake, with an area of just over 1,300 square miles, followed by Okeechobee (700), Pontchartrain (625), Champlain (435), and Tahoe (193).*

⑩ *Ontario, Erie, Huron, Michigan, and Superior. The largest is Lake Ontario.*

⑪ *Lake Geneva.*

⑫ *Canada. The great number of lakes is largely the result of glacial action during the Pleistocene period (Ice Age) when they were gouged in the earth.*

⑬ *The Lost Sea.*

⑭ *In southwestern Alberta, Canada, in the Rockies.*

⑮ *They are eyeless, as are many animals and insects that dwell in the total darkness of caves.*

⑯ *Walden Pond, the setting and title of Henry David Thoreau's book, published in 1854, urging people to get back to nature and live simpler, more meaningful, lives, is located in Concord, Massachusetts.*

⑰ *Crater Lake is constantly replenished by rain and snow, and through plant and animal life, maintains a proper balance without significant deterioration.*

(18) *The northern states have far more lakes than the southern ones, mainly because the glaciers of the Ice Age, which gouged out natural lake beds and at the same time filled them with water from the melting ice, did not reach the lower states.*

(19) *Lake Placid is in northeastern New York. It was selected as the site for the 1938 and 1986, as well as the 1932 Winter Olympics because of its location in the Adirondacks, surrounded by high mountains that are ideal for skiing, bobsled runs, and other winter sports.*

(20) *Minnesota, which has 4,854 square miles of water. Florida, with 4,511 square miles, has less water but a greater* number *of lakes, many of them very small.*

(21) *(d) The Soviet Union, which contains this huge swamp in a basin of the Pripyat River. The swamp is the heart of the Pripyat Marshes, which include some 38,000 square miles of forests and swamplands.*

(22) *Ghana, on the southern coast of West Africa.*

(23) *The countries with shorelines on Lake Tanganyika are Burundi, Tanzania, Zaire, and Zambia.*

(24) *A* mere *is another term (chiefly British) for a pond or small lake.*

㉕ *1(d) New York/Central Park Lake, 2(d) Washington/ Tidal Basin, 3(b) Boston/Lake on the Commons, 4(c) Miami/Blue Lagoon Lake, and 5(a) Dallas/White Rock Lake.*

㉖ *Minneapolis, which, like its state, Minnesota, is speckled with lakes large and small. It has no fewer than thirty lakes within the city limits, not to mention dozens of ponds and other small bodies of water.*

㉗ *The Po, which is 405 miles long, is the longest river in Italy.*

㉘ *It empties into the Tyrrhenian Sea. The city is Rome.*

㉙ *(a) The Seine. (b) It flows northwest from its source in Burgundy and empties into the English Channel at Le Havre.*

㉚ *The south side. Left and right banks of a river are determined by facing downstream.*

㉛ *The River is the Tiber, the city is Rome, and the country is the Vatican.*

㉜ *The Danube, because it was known to the ancient Greeks and many early civilizations along its 1,770-mile course from its source in the Black Forest through Germany, Austria, Hungary, Romania, and Bulgaria to the Black Sea.*

③③ *The Volga River, which never leaves the Soviet Union during its 2,300-mile course.*

③④ *(a) The Thames, (b) pronounced "tems," rhyming with "gems." (c) It flows into the southern reaches of the North Sea, just north of the English Channel.*

③⑤ *The Tagus, which has its origin in Spain, flows westward to empty into the Atlantic Ocean.*

③⑥ *The Continental Divide, known as the "backbone" of North America, is the great ridge of the Rocky Mountains, sometimes referred to as "The Great Divide," which separates westward-flowing streams from eastward-flowing waters. It runs from northern Alaska to southwestern New Mexico.*

③⑦ *The Ohio River, which joins at Cairo, Illinois, and the Missouri River, which joins at St. Louis, Missouri.*

③⑧ *The Yukon River, named because of its origin in Canada's Yukon Territory.*

③⑨ *(a) The St. Lawrence River. (b) It runs through Montreal and Quebec.*

④⓪ *(a) The Potomac originates in the Allegheny Mountains. (b) It flows into the Chesapeake Bay.*

④① *The Tennessee River.*

④② *Burma, Cambodia, Laos, Thailand, and Vietnam.*

④③ *The Rio Grande.*

④④ *The Murray and the Darling, both in the southeastern region of the country, are the only major rivers in Australia.*

④⑤ *The Nile (4,160 miles). The second longest is the Amazon (3,990 miles).*

④⑥ *(a) The Hwang Ho (Yellow River) and the Yangtze. (b) The latter empties into the East China Sea at Shanghai.*

④⑦ *The Vistula River, whose mouth is near Elblag on the Gulf of Danzig.*

④⑧ *The Missouri River, which originates in southwestern Montana and flows into the Mississippi.*

④⑨ *The Ohio River, formed in the heart of Pittsburgh, Pennsylvania, by the junction of the Monongahela and Allegheny Rivers.*

⑤⓪ *(a) The James River. (b) Jamestown was built there some fifty miles inland from the Atlantic coast.*

⑤① *The Congo River. The wide variety of life in and around its waters is the natural consequence of exposure to the multitudes of environments along its course from the mountains to the sea.*

(52) *Texas to the south of the Red River, and Oklahoma to the north.*

(53) *Stratford-on-Avon, appropriately named because it is on the Avon River, which flows for less than 100 miles through the heart of England.*

(54) *The Hudson River.*

(55) *(a) The Somme is a river in Picardy, France. (b) It was the scene of four furious and bloody battles during World War I.*

(56) *(a) The Yalu is in North Korea, on the Manchurian border. (b) It was a strategic location during the Korean War (1950–53).*

(57) *(a) The Blue Nile, which starts in the Ethiopian highlands, and the White Nile, which flows out of Lake Victoria in Uganda. (b) They meet at Khartoum, in the Sudan.*

(58) *(a) The Rio Grande. (b) It is 1,800 miles long.*

(59) *The Volga-Baltic Canal in Russia, which is more than 1,800 miles long, running from the Caspian Sea to Leningrad.*

(60) *A canal lock is an enclosed section of water, with gates at each end, for raising or lowering vessels from one level to another*

*by admitting or releasing water. Locks are necessary when the bodies of water connected by a canal are at different sea levels.*

(61) *The Panama Canal, with Gatun Lake.*

(62) *Along part of its course it disappears underground, only to appear again later. There are other lost rivers in the world that disappear below the surface for all or part of their courses.*

(63) *The Niger, in Nigeria. The origins of the name are obscure, but probably stemmed from the fact that the river was muddy in color and (like many in Africa) was worshipped as a source of food and an avenue of transportation.*

(64) *The Amazon, which pours out so much water at its mouth that it literally pushes the surrounding salt water out to sea.*

(65) *Montreal, which is situated on Montreal Island. It was located there when Montreal was a trading post to make access more difficult by enemies, and later was fortified.*

(66) *The Dead Sea, which is the lowest point on earth and some 1,300 feet below the level of the nearby Mediterranean.*

(67) *Iberia is a peninsula. The countries are Portugal and Spain, separated from the rest of Europe by the Pyrenees.*

**68** *1 (d) Lena/Soviet Union, 2(e) Orinoco/Venezuela, 3(a) Madeira/ Brazil, 4(c) Mackenzie/Canada, and 5(b) Euphrates/Turkey.*

# Islands

**W**hy are people so intrigued with the islands of the world? Possibly because each island, like each person, is an entity unto itself. Depending upon their size and nature, islands are ecosystems that may also contain a wide variety of other geographical features, such as mountains, lakes, rivers, peninsulas, plains, glaciers, volcanoes, canyons, and even deserts. No wonder that islands have long been the subject of so many multitudes of books, articles, songs, and paintings. No wonder, too that they have played their part in the history, economy, and politics of the world.

Let's go island-hopping!

**①** We start with an island whose popularity has been unquestioned since it was first visited by the Spanish in 1515. It is not just one island, but a collection of some 300 coral rocks, islets, and islands in the Atlantic, southeast of Cape Hatteras, North Carolina. Can you name it?

**②** Match these islands with the nations that own them:

| Island | Nation |
|--------|--------|
| **1.** Spitsbergen | **(a)** France |
| **2.** Corfu | **(b)** Indonesia |
| **3.** Kodiak | **(c)** Norway |
| **4.** Java | **(d)** Greece |
| **5.** Martinique | **(e)** United States |

**③** What is the group name for these islands: Alderney, Guernsey, Jersey, and Sark?

**④** In what body of water are the islands in the preceding question located?

**⑤** This island in the Marianas Islands of the Pacific is a self-governing United States territory whose slogan is "Where America's day begins." What is the name of this island?

**⑥** Many of the islands in the Pacific, including those of Hawaii, are recognized as islands that rose directly from the ocean floor without ever having been attached to a continental mainland. How did islands of this type originate?

**7** The island of Trinidad is a type known specifically as a "continental island" because it was originally joined to a continent and still has many species of animals and birds that can be found on the land from which it separated. Which continent are we referring to?

**8** At the tip of Florida lies a string of islands stretching out into the Gulf of Mexico and much visited by tourists. Can you name them?

**9** Referring to the previous question, what does the term mean?

**10** What do these islands have in common: Hokkaido, Honshu, Kyushu, and Shikoku?

**11** Exactly where are the above islands located?

**12** Certain islands form a buffer between the open ocean and the mainland shore. They guard some thirteen percent of the world's ocean shoreline and, characteristically, have beaches on the ocean side, sand dunes down their spines, and shallow, marshy areas on the mainland side. What is this type of island called?

**13** What do these islands have in common: Leyte, Luzon, Mindanao, and Panay?

**14** The Thousand Islands group contains more than 1,800 islands, all formed at the end of the Ice Age. Where is this group located?

**15** This large island in the Mediterranean is roughly triangular in shape and lies close to the southwestern tip of Italy, separated only by a narrow strait. Can you name the island? For extra credit, what is the strait?

**16** Here's a two-parter: The Republic of China, which broke off from the People's Republic of China (mainland China) in 1949 and set up a new government under General Chiang Kai-shek, lies almost wholly on an island about twice the size of New Jersey. (a) What is the name of the island, and (b) what is the body of water that separates it from the mainland?

**17** The Isle of Wight, with its mild climate and many scenic attractions, including cliffs and beaches that are composed of vari-colored sands, is a popular resort area. Where would you travel to if you wanted to vacation on the Isle of Wight?

**18** Some islands become known for a single prominent feature or form of life. Chincoteague and Assateague, narrow barrier islands off the coast of Maryland and Virginia, are good examples. For what are they noted: (a) wild ponies, (b) sea snakes that lay eggs on their beaches, or (c) stands of dune bamboo used for making furniture?

**19** There are several countries in Asia that are made up of geographical entities known as archipelagos. What is an archipelago?

**20** Referring to the preceding question, what are two examples of major archipelagoes?

**21** The Sunda Islands of Indonesia, lying between the South China Sea and the Indian Ocean in the Malay Archipelago, are noted for being the home of a very rare beast, the Komodo Dragon. What exactly is a Komodo Dragon?

**22** New Yorkers know Long Island, the largest island in the continental United States, as the site of beaches, suburban residential communities, summer colonies, and (unfortunately) frequent traffic jams. If you were to drive from the New York City end 118 miles to the very tip of Long Island, where would you be?

**23** Name at least three of the tropical or semitropical islands in the Greater Antilles or the Lesser Antilles of the West Indies.

**24** These major ports in the United States have islands that are tiny, but historically significant, in their harbors. Match the cities with their islands:

| City | Island |
|------|--------|
| **1.** New York | **(a)** Angel Island |
| **2.** Miami | **(b)** Sullivans Island |
| **3.** Baltimore | **(c)** Governor's Island |
| **4.** San Francisco | **(d)** Key Biscayne |
| **5.** Charleston | **(e)** Fort McHenry |

**25** Very few people realize that New Hampshire has a seacoast on the Atlantic Ocean. Although only ten miles long, it is located at the city of Portsmouth. What group of islands with a hazardous-sounding name lies off this coast?

**26** The Isles of the Blessed (also called the Fortunate Isles) were mentioned in classic mythology as islands where the souls of favored mortals were received by the gods and lived happily in paradise. They were said to be identified with the Canary Islands. Where are the Canary Islands?

**27** Manitoulin Island in northern Lake Huron is eighty miles long and from two to thirty miles wide. It is unique in two ways, one relating to its size and the other to bodies of water on the island. Can you tell why?

**28** A major military base for the United States Marines is Parris Island. Where is it?

**29** The third largest island in the world is in the Malay Archipelago. It lies south of the Philippines and east of Singapore. Name it.

**30** Lying just to the south of the island in the preceding question is an island that has one of the densest populations in the world, is very cosmopolitan, and is the center of much Asian culture. Can you name it? For extra credit, name both the country it belongs to and the capital of that nation, which happens to be located on the island in question.

**31** Islands are born in a number of different ways, usually requiring thousands, even millions, of years to evolve into their present form. The island of Surtsey, which lies a few miles southwest of Iceland in the North Atlantic, is a dramatic exception. In early November 1963 the site was open ocean. By February of the following year, Surtsey was an island covering one square mile and rising 670 feet. Surtsey is an example of what kind of island?

**32** Christmas Island was discovered in 1777 by Captain James Cook when he was exploring the Pacific Ocean. He is most associated, though, with a group of islands that now belong to the United States and lie about 1,000 miles north of Christmas Island. Can you name these islands?

**③③** The Hebrides are a group of about 500 islands. Where are they located?

**③④** The Fiji Islands are really the Viti Islands, but American missionaries mispronounced the name and it stuck. There are about 250 of these islands lying in the South Pacific. Which continent is closest to them?

**③⑤** Australia has very few inhabited islands. The largest and by far the most important lies off the southeastern tip of the continent, has several major cities, and has a land area of more than 26,000 square miles (about equal in size to West Virginia). For a two-parter: name (a) the island, and (b) its capital.

**③⑥** South of Turkey lies an island about 3,500 square miles in size, where the people speak both Turkish and Greek. What is this island?

**③⑦** West of Morocco in the Atlantic Ocean lie two groups of islands, the Canary Islands and Madeira. To which country does each group of islands belong?

**③⑧** In considering the various continents and regions of the world, what does the term "Oceania" mean from a geographic standpoint?

**39** Melanesia, a major group of islands in the Pacific Ocean, includes Papua New Guinea, the Solomon Islands, and Fiji. Many U.S. Marines who served in the Pacific during World War II became more familiar with the Solomons than they ever cared to be because of the fierce fighting there. Where are the Solomons?

**40** Micronesia, meaning "little islands," is another major group of islands in the Pacific. What three island chains are part of Micronesia?

**41** A third group of Pacific islands is Polynesia, which extends from New Zealand in the southwest to Easter Island in the southeast, and Hawaii in the north. What are the two major components of New Zealand?

**42** What is the capital of New Zealand, and where is it located?

**43** Most people are thoroughly familiar with Hawaii, formerly a kingdom of its own, from articles and films if not from personal visits. But can you name at least half of the eight islands that comprise this Pacific archipelago?

**44** The Shetland Islands are perhaps best recognized as the home of the Shetland pony, the smallest breed of horse. The islands are also of great geological and archaeological interest because of their structure and ability to sustain an ancient civilization. What nation owns them?

**45** The highest island mountain in the world, measured at 32,000 feet from base to peak, has been the subject of geological investigation for more than 150 years because of its origins and dramatic outbursts as an active volcano. What is it?

**46** These islands are the subject of much study today by geologists who are interested in the lava piles that characterize much of the land, and by zoologists who camp there to study the families of tortoises, iguanas, and many sea birds who are prevalent inhabitants. For a two-parter: (a) what is the name of these islands, and (b) what was the name of the noted scientist who visited them more than a century and a half ago to determine how certain species evolved?

**47** Which one of the following nations lies entirely by itself on an island: (a) Surinam, (b) Laos, (c) Iceland, or (d) Botswana?

**48** Mont-Saint-Michel is an historic island, which is accessible by land at low tide. It contains a celebrated Benedictine abbey that was founded in 708 and today consists of a massive group of buildings, including a great abbey church. Where is Mont-Saint-Michel?

**49** These islands are bleak and rocky, swept by wind and drenched by chilling rains most of the year. They are a British dependency and lie 300 miles east of the Strait of Magellan at the southern tip of South America. What is the name of this group?

**50** Much more pleasurable than the islands described in the previous question, these are tropical, bathed in sun most of the year, and lush with vegetation. The islands and islets, some 700 in all, form an independent commonwealth and lie between the southern tip of Florida and Haiti. What is the name of this group?

**51** This island, which had one of the world's earliest civilizations, is the largest of the Greek islands (160 miles long) and lies in the Mediterranean, about 60 miles off the mainland of Greece. What is its name? For extra credit, what was the name of this early civilization?

## Answers

**①** *Bermuda.*

**②** *1(c)  Spitsbergen/Norway, 2(d) Corfu/Greece, 3(e) Kodiak/United States, 4(b) Java/Indonesia, and 5(a) Martinique/France.*

**③** *The Channel Islands.*

④ *They are in the English Channel between France and England.*

⑤ *Guam. The slogan derives from the fact that the island lies to the west of the international dateline. Thus, when it is Tuesday in the continental United States, the sun has already risen for Wednesday in Guam.*

⑥ *Most such islands orginated as volcanoes.*

⑦ *Trinidad was once joined to the northeastern coast of South America at a point in Venezuela where the Orinoco River flows into the Caribbean Sea.*

⑧ *They are the Florida Keys, the largest of which is Key Largo, about thirty miles long.*

⑨ *A key (sometimes called "cay") is a small, low island characteristically made of sand or coral.*

⑩ *They are the major islands that make up Japan.*

⑪ *They are located between the North Pacific Ocean and the Sea of Japan between latitudes 30°N and 40°N.*

⑫ *A barrier island. Hilton Head, South Carolina, is one of the largest barrier islands on the east coast of North America.*

⑬ *These are four of the eleven largest islands that make up the Republic of the Philippines. They are located at the western end of the Pacific Ocean, between the Philippine Sea and the South China Sea, just north of the equator.*

⑭ *The Thousand Islands group lies in the St. Lawrence River east of Lake Ontario and north of the state of New York.*

⑮ *The island is Sicily. The narrow waterway separating it from Italy is the Strait of Messina.*

⑯ *(a) Taiwan (Formosa). (b) It is separated from the mainland by the Strait of Formosa.*

⑰ *You would travel to the English Channel, just off the south central coast of England.*

⑱ *(a) Chincoteague and nearby Assateague are the homes of wild Chincoteague ponies, descendants of animals that swam ashore from a wrecked ship in the sixteenth century.*

⑲ Archipelago *was the ancient name for the Aegean Sea and later was used to designate the numerous islands in that sea. Today it refers to any cluster of islands.*

⑳ *There are dozens of large archipelagos in the seas of the world. Among* the most familiar are: the Philippines, Indonesia, Hawaii, the Kurils, and the Aleutians.

㉑ The Komodo Dragon is the world's largest lizard. It may grow to as much as ten feet long and attain a weight of more than 300 pounds. One of the Sunda islands, Rinja, has been set apart as the world's only dragon sanctuary.

㉒ Montauk Point.

㉓ Cuba, Jamaica, Haiti, the Dominican Republic, Puerto Rico, the Leeward Islands, the Windward Islands, Trinidad and Tobago, and Barbados.

㉔ 1(c) New York/Governor's Island, 2(d) Miami/Key Biscayne, 3(e) Baltimore/Fort McHenry, 4(a) San Francisco/Angel Island, and 5(b) Charleston/Sullivans Island.

㉕ The Isles of Shoals, a rocky and rugged group of small islands and reefs.

㉖ The Canary Islands are located not far west of the Mediterranean off the western coast of Morocco.

㉗ Manitoulin is the world's largest lake island and within its borders contains more than 100 lakes and ponds.

㉘ Parris Island is located near the city of Beaufort, South Carolina, on the Atlantic coast.

㉙ *Borneo. Despite its size, it is sparsely populated, partly because it sits smack on the equator, has an annual rainfall of more than one hundred inches, and has a surface that consists of swamps, jungles, high peaks, and other terrain hostile to humans.*

㉚ *Java is the name of the island. It belongs to Indonesia, whose capital and largest city, Djakarta (formerly Batavia), which has a population of almost nine million people, is located on Java.*

㉛ *Surtsey is an example of a volcanic island, also known as an oceanic island. In 1963 volcanic action forced the undersea cone to rise above the surface, steaming and smoking. The event was recorded on film from the air and sea. Today the island is the habitat of numerous birds. Biologists have observed the way in which the island has "taken root" and started to support plant life as well.*

㉜ *The Hawaiian Islands.*

㉝ *The Hebrides, which include the islands of Skye, Mull, and Iona, are off the west coast of Scotland.*

㉞ *Australia.*

㉟ *Tasmania, (b) Hobart.*

㊱ *Cyprus. The island is an independent republic whose population is com-*

*posed seventy-eight percent of people of Greek descent and about nineteen percent of people with Turkish origins.*

(37) *The Canary islands belong to Spain, and Madeira to Portugal.*

(38) *Oceania is the collective name applied to the approximately 25,000 islands of the Pacific, including Australia but excluding Indonesia, the Philippines, and others whose cultures are more closely related to the mainland. Many of these islands are unnamed, unpopulated, and too barren for cultivation.*

(39) *The Solomons are a string of islands in the Coral Sea, northeast of Australia and east of New Guinea.*

(40) *The Carolines, the Marianas, and the Marshall Islands.*

(41) *New Zealand is comprised of two large areas: North Island, which is the smaller but has about seventy percent of the population, and South Island, which is noted for its massive and glorious Alps.*

(42) *The capital is Wellington, located on North Island.*

(43) *From north to south they are: Kauai, Niihau, Oahu, Molokai, Maui, Lanai, Kahoolawe, and Hawaii.*

**44** Scotland. In ancient times, the Shetlands were inhabited by Norsemen, but were annexed to Scotland in 1472.

**45** Mauna Kea, in Hawaii. Its enormous height (greater than Everest) is measured from the bottom of the ocean floor, which lies more than 18,000 feet below sea level.

**46** (a) The Galapagos. (b) Charles Darwin, who sailed to the islands in his famed ship, Beagle, in 1835, to gather data to support his theory of evolution.

**47** (c) Iceland.

**48** In the Gulf of Saint-Malo, an arm of the English Channel, in northwestern France.

**49** The Falkland Islands.

**50** The Bahama Islands.

**51** Crete. It was the site of the Minoan civilization which flourished as far back as 3000 B.C.

# Deserts and
# Polar Regions

This book would be incomplete if it did not ask some basic questions about the remote deserts, polar regions, and other so-called wastelands of our planet, and provide a few answers and pertinent descriptions. So it's time to settle back in a comfortable seat and be prepared to take an imaginary tour of some places that you would not usually select for a pleasure trip.

Let's start first with some of the hotter places on earth.

**❶** Deserts are noted for being inhabited by some very strange animals. Quite a few in the rabbit and rodent families, for instance, have unusually large ears, not to hear better with, but to perform a function related to the environment. What is this function?

**2** Only one continent does not have any deserts. Which continent is this?

**3** The very word "desert" connotes a vast sea of sand to most of us. But how much of the world's deserts is actually sand: (a) 20%, (b) 42%, or (c) 78%?

**4** One of the world's highest deserts is in central Asia, extending east and west across southeastern Mongolia and northern China. It is about one thousand miles in length, lies on a plateau at altitudes ranging from 3,000 to 5,000 feet, and is also one of the coldest deserts, having long winters and only short summers. Name it.

**5** What is the world's largest desert, and where is it located?

**6** By definition, deserts are dry, with little rainfall and only sparse vegetation. Nevertheless, one desert easily holds the record as being the driest of all. Can you name it?

**7** One of the most familiar deserts in the United States is in southern California. It is small in comparison with the major deserts of the world, only 15,000 square miles. What is it?

**8** Rub al Khali ("empty quarter") is sixth in size in the world and is noted for its giant sand dunes, some of which rise to heights of more than 600 feet. Where is this desert?

**9** This 200-mile stretch of "badlands" in northeastern Arizona near the Grand Canyon is famous for its striking bands of color that result from irregularly eroded layers of red and yellow sediments and various shades of clay. Name it.

**10** What is the second largest desert in the world?

**11** Far from being total wastelands, some desert regions are greatly desired by the people who inhabit them. One of the most hotly contested deserts is in southern Israel and has frequently been the scene of battles between opposing forces who claim it. (a) What is its name, and (b) why is it so sought after?

**12** How much of the world's surface is covered by true deserts: (a) one tenth, (b) one sixth, or (c) more than one-fourth?

**13** Where would you find the third largest desert in the world?

**14** This desert, in Botswana, South Africa, is the seventh largest. It is somewhat different from other deserts as it is studded with dry lake beds that collect enough moisture during the rainy season to support sparse amounts of grass and scrub. What is its name?

**⑮** A British sports writer boasted that he had caught fish all over the world, including several in the Sahara Desert. Was he pulling the reader's leg?

**⑯** The Atacama Desert in Chile is, despite its total lack of rainfall, very important to the economy of this South American republic. Why?

**⑰** Simooms and Haboobs are desert phenomena, which make those wasteland environments even more unpleasant and treacherous than usual. What are they?

**⑱** If you were traveling across a desert in the Middle East, you would be likely to encounter certain native peoples called: (a) Malabars, (b) Sherpas, (c) Bantus, or (d) Bedouins?

**⑲** You probably know one of the main reasons why camels are used for carrying people and their belongings across deserts: they can survive for days at a time with no water. But camels have other characteristics that make them eminently suited for desert travel. Can you name them?

**⑳** Armies that engage in desert warfare have developed tanks and other mechanized equipment with which to fight the enemy. What are the three major obstacles designers have had to contend with in producing war machines that would work efficiently?

**21** There is a strange phenomenon in the desert, which has been described time and again as "singing dunes." Travelers have reported the sound as being loud enough to be heard almost a mile away and sometimes continuing for five or ten minutes. Can you guess what might cause the desert sands to "give tongue"?

**22** Deserts in the western United States are noted for their species of cactus, which range from the size of a pea to a species that reaches heights of fifty feet or more. What is the name for this hardy desert giant? (Hint: Its blossom is the state flower of Arizona and it is the name for a national monument in southeastern Arizona.)

**23** This desert has the distinction of being the lowest point in North America (282 feet below sea level) and having the hottest temperatures (134°F). Can you name it?

**24** Another famous desert in the United States was once an enormous lake. It still contains a tiny portion of that lake, dwindled down to a fraction of its former size in the desert's 1,500-square-mile area. What is the name of this desert and the lake it contains?

**Now let's see how hot you are on cold questions. If you find it difficult to answer questions about these remote regions of snow and ice, you are not alone. Even today, aided by the use of the most sophisticated methods and the most advanced instruments, geographers and geologists find many aspects of the polar world perplexing, if not contradictory.**

**How well do you function at 50° below zero?**

**㉕** One of the continents is Antarctica, the fifth largest in the world, which covers some 5,500,000 square miles around the South Pole. But there is no Arctic continent at the North Pole. Why is that?

**㉖** Scientists and researchers studying the Antarctic often have to make deep tunnels to reach the land formations that lie beneath the frozen surface. What is the average thickness of the ice sheet that covers most of the Antarctic: (a) about 1,000 feet, (b) 3,200 feet, or (c) more than a mile?

**㉗** On certain clear nights, electrical solar discharges cause flashing lights in the upper atmospheres of both polar regions, which can be seen for hundreds of miles away. What is the name for those in the region of (a) the North Pole, and (b) the South Pole?

**28** Since the North Pole and the South Pole are each situated at about the same distance from the equator, their climate and temperatures should be about the same during the respective seasons of the year. Is this true?

**29** Explorers crossing polar regions are in danger of losing their way because of a phenomenon called "white-out." Just what is this hazard?

**30** The northern tip of what large island is the land nearest to the North Pole?

**31** Arctic seals are mammals and thus have to breathe every seven to nine minutes, or up to twenty minutes in an emergency. How, then, can they survive under the ice pack, catching fish and living without ever kicking their way up onto the ice?

**32** Intimately associated with the Arctic is the tundra, a region that stretches north as far as the shores of the Arctic Ocean. Is the tundra: (a) an uneven layer of ice that encrusts the land, (b) vast treeless plains that contain a variety of hardy grasses and scrub, or (c) forests of stunted evergreens that survive in sub-zero temperatures?

**33** By comparison with the United States, how big is Antarctica?

**34** The arctic regions could never have been explored in the days of polar discoveries without the use of an age-old Eskimo method of transportation. Describe it.

**35** What do the following names have in common: Ross, Weddell, Bellinghausen, and Amundsen?

**36** One of the many hidden dangers that threaten explorers in their journeys across snow and ice is a large crack in the surface that may extend downward from thirty feet to more than one hundred feet. What is the name of this hazard?

**37** This archipelago belongs to Alaska, is made up of rugged volcanic islands, and has a cold, wet subarctic climate. Can you name this island chain?

**38** How much do you know about our forty-ninth, and coldest, state, Alaska? Try this matching question:

| City | Fact |
|------|------|
| **1.** Nome | **(a)** Capital |
| **2.** Juneau | **(b)** Largest population |
| **3.** Anchorage | **(c)** Farthest west |

**39** Ketchikan, a fishing center and, historically, a supply point for miners during the gold rush of the 1890s, is both the southernmost and easternmost city in Alaska. If you were to fly directly east, would you pass over any other cities in the United States? What if you flew south?

**40** At what latitude (approximately) does the Arctic Circle lie?

**41** The white bears found in icy regions within the Arctic Circle are known as polar bears. Do similar bears also live in the Antarctic?

**42** Kayaks were—and still are today—one of the most important modes of transportation for Eskimos traveling amongst ice floes in the north polar regions. What were these made of?

**43** Before Alaska became part of the United States, what country did it belong to?

**44** Because Alaska was long considered a wasteland, it was given a derisive nickname at the time of its purchase. What was it called?

**45** Adjoining Alaska to the east is an area in Canada that lies, in part, north of the Arctic Circle. It became famous in the 1890s when more than 30,000 people pushed across its icy barriers in search of gold. What is the name of this area?

**46** There are three major currents in the Arctic, one of which is the Greenland Current which circles most of the island of Greenland. Into what two oceans does this current flow?

**47** What is the name of the 700-mile-long, ice-clogged body of water that runs along the west coast of Greenland?

**48** At what latitude (approximately) does the Antarctic Circle lie?

**49** Admiral Richard E. Byrd was a famous American flier and explorer who commanded expeditions to both the North and South Poles. In 1929 he led a well-equipped expedition to Antarctica where he established a base on the Ross Ice Shelf that was much publicized in the United States. What was the name of that base?

**50** Name three or more countries that lie in part within the Arctic Circle.

Try this True/False test about the two polar regions:

**51** True or false. The Antarctic ice cap is so immense that, if it were melted, it would produce more fresh water than in all the rest of the world.

**52** True or false. Penguins similar to those in the Antarctic also live in the Arctic.

**53** True or false. The summer temperature in the Arctic can rise as high as 50°F.

**54** True or false. The summer temperature in the Antarctic can frequently climb to 40°F.

**55** True or false. Less than five percent of Antarctica is free of ice.

**56** Match these famous explorers with the countries they represented during notable expeditions to the Arctic:

| Explorer | Country |
|---|---|
| **1.** Roald Amundsen | **(a)** United States |
| **2.** Umberto Nobile | **(b)** Sweden |
| **3.** Robert E. Peary | **(c)** Norway |
| **4.** Salomon Andree | **(d)** Canada |
| **5.** Vilhjalmur Stefansson | **(e)** Italy |

## Answers

① *In case you have never noticed, the ears of humans and animals are crisscrossed with many small blood vessels. Breezes, even warm ones, cool these vessels and, in turn, the blood itself. Because the ears of these desert animals are so large, they have much greater cooling power than normal-sized ears.*

② *Europe is the only continent that has no desert.*

③ *(a) Only one-fifth of the average desert is sand. The rest is made up of rock outcroppings, loose gravel, or pebbles.*

④ *The Gobi Desert.*

⑤ *The Sahara, which covers 3.5 million square miles. It is in North Africa.*

⑥ *The Atacama Desert, which comprises most of the entire land mass of Chile in South America, sets the record. It contains almost no vegetation and has areas where no rainfall has ever been recorded.*

⑦ *The Mojave. It joins the Colorado Desert in the southeast.*

⑧ *On the Arabian Peninsula (called Arabistan in Persian). The Rub al Khali Desert is considered part of Saudi Arabia, but it is not really claimed by anyone since it is waterless and cannot support any form of life.*

⑨ *The Painted Desert.*

⑩ *The Libyan Desert, east of the Sahara, which contains 650,000 square miles.*

⑪ *(a) The Negev, more than 5,000 square miles in area. It is bordered by the Judaean Highlands and the Sinai Peninsula. (b) It represents half of Israel's land area and is valuable because part of it can be irrigated economically for crops and it is rich in natural gas and mineral deposits.*

⑫ *(c) About one-fifth of the world's surface is made up of hot deserts and one-sixth cold deserts.*

⑬ *The world's third largest desert is the Great Australian Desert, in central and western Australia.*

⑭ *The Kalahari Desert.*

⑮ *Probably not. The Sahara has a number of* wadis *(intermittent streams) that run largely beneath the surface but that do contain small fish and other aquatic life. It is doubtful, however, that such fish are large enough or meaty enough to provide a very tasty meal.*

⑯ *The Atacama Desert contains rich deposits of copper, as well as nitrates for the production of fertilizers.*

⑰ *Simooms and haboobs are violent whirlwinds that race through the deserts, sometimes with leading edges that appear as solid walls of dust as much as 5,000 feet high. These storms in North Africa and Arabia are largely responsible for the atmospheric dust over Europe.*

⑱ *(d) Bedouins. They are nomad Arabs who raise sheep, use camels for transportation, and are family tribes headed by a sheik.*

⑲ *They store fats, as well as liquids, in their humps, have thick, broad hoofs that do not sink into the sand, and are blessed with nostril hairs that filter out sand and dust when they breathe.*

⑳ *1. Extreme heat, 2. constant infiltration of dust and sand particles, and 3. lack of readily available water supplies.*

㉑ *When winds pile up the sand in a dune too steeply, the tiny grains start to slide downward, and when they do so in enormous quantities, they rub against each other and set up a humming noise similar to that of a stringed instrument or the whirring of a top.*

㉒ *The saguaro, many of which can be found in the Saguaro National Monument.*

㉓ *Death Valley, one of the most famous deserts in the United States, located in southeastern California and southwestern Nevada.*

㉔ *Great Salt Lake Desert and Great Salt Lake.*

㉕ *There is no Arctic continent because the center of the Arctic region where the North Pole is located is all water, an immense basin that comprises the Arctic Ocean. The north polar ice cap and the vast regions of solid snow and ice are simply floating on this sea.*

㉖ *(c) All but the barest edge of the continent is encased in this crust that is one mile thick or more.*

㉗ *(a) The Aurora Borealis, or Northern Lights, at the North Pole, and (b) the Aurora Australis at the South Pole.*

㉘ *No. The South Pole has by far the more severe climate, mainly because of its altitude, which rises in some places to almost 15,000 feet. Temperatures of 140° below freezing have been recorded at the South Pole—50° colder than the lowest at the North Pole, which is almost at sea level.*

㉙ *When there are low-lying clouds overhead, light rays may bounce from them to the snow and ice below, creating a blinding opalescence, a white-out, that makes it impossible for a person to see anything but white or get a bearing in any direction.*

㉚ *Greenland.*

㉛ *After going under the pack, seals chew "chimneys" up through the ice that serve as breathing holes. A series of such passages is kept open in the area where the seals are swimming and foraging.*

㉜ *(b) The tundra is a region of some five million square miles of wide plains and frozen subsoil that stretches around the shores and islands of the North Sea.*

㉝ *Antarctica, the Antarctic continent, is one and a half times the size of the United States.*

㉞ *Sleds pulled by Eskimo dogs or huskies.*

㉟ *They were all explorers who made significant discoveries in the Antarctic and had seas named after them.*

㊱ *A crevasse or crevice. What makes it doubly dangerous is that it is usually hidden, covered with a snow bridge which can give way under the weight of one or more people walking or skiing along the surface.*

㊲ *The Aleutian Islands, which stretch westward like a tail from the southern shores of Alaska.*

㊳ *1(c) Nome/farthest west, 2(a) Juneau/capital, and 3(b) Anchorage/largest population.*

㊴ *You would not pass over any other American cities because even this point in Alaska is too far north and too far west.*

㊵ *The Arctic Circle is an imaginary circle on the face of the globe at 66½° north latitude, or 23½° south of the North Pole. It marks the northernmost point at which the sun can be seen at the winter solstice (about December 22).*

㊶ *There are no polar bears, or for that matter any other kinds of bears, in the Antarctic.*

㊷ *Sealskins, stretched over frames of wood or bone.*

**43** *Russia. The Russians originally explored the region, but by the middle of the nineteenth century their interest declined and the land that is now Alaska was sold to the United States in 1867 for $7,200,000.*

**44** *Alaska was referred to as "Seward's Icebox" or "Seward's Folly" because its purchase in 1867 was effected largely through the efforts of William H. Seward, who was then secretary of state.*

**45** *It is the Yukon Territory. Canadian explorers were told by the Indians that this was* yukon-na, *meaning "a big river," and the name stuck.*

**46** *The Greenland Current flows into the Arctic Ocean and the North Atlantic.*

**47** *Baffin Bay.*

**48** *The Antarctic Circle is an imaginary circle on the face of the globe at 66½° south latitude, or 23½° north of the South Pole. It marks the southernmost point at which the sun can be seen at the summer solstice (about June 22).*

**49** *Little America.*

**50** *The countries that lie in part within the Arctic Circle are: the Soviet Union, Finland, Sweden, Norway, Iceland, Greenland, Canada, and the United States.*

ⓢ *True.*

ⓢ *False.*

ⓢ *True.*

ⓢ *False.*

ⓢ *True.*

ⓢ *1(c) Amundsen/Nor-way, 2(e) Nobile/Italy, 3(a) Peary/United States, 4(b) Andree/Sweden, and 5(d) Stefansson/Canada.*

# Peoples and Populations

We tend to think of geography as a science related only to the earth and its waters and land features. Yet a significant component is "human geography," which covers the arrangement and distribution of people in every corner of the globe. Human geographers (sometimes called population geographers) focus their efforts on analyzing the composition of the population of given places, according to such classifications as age groups, races, religions, languages, and sexes. This seems logical and appropriate for this particular science when you consider that the nature of the earth's waters and landforms play a vital part in determining where people will live and how they *will develop their skills and cultures, among other things.*

*Political geography is a subdivision of human geography that not only studies peoples as individuals and families, but analyzes the ways in which they are organized and grouped as communities, states, and nations.*

*So, with this in mind, and thinking about the physical aspects of geography and environment that influence where people live and what they do, let's see how well you know your world of people and populations.*

*First, let's sharpen our image of the vast numbers of people we are dealing with when we think of populations.*

❶ Approximately how many people are there in the United States today: (a) 25,000,000, (b) 250,000,000, or (c) 2,500,000,000?

**❷** What is the total population of the earth: (a) 500,000,000, (b) 5,000,000,000, or (c) 50,000,000,000?

**❸** How many people do you think lived on earth in 1950: (a) 25,000,000, (b) 250,000,000, or (c) 2,500,000,000?

**❹** The Population Division of the United Nations not only computes current populations, country by country, but undertakes to make projections for the future. What figure has this division projected for the year 2000: (a) 5,800,000,000, (b) 6,127,000,000, or (c) 9,355,000,000?

**❺** What percent of United States citizens are immigrants or descendents of immigrants: (a) twenty-five percent, (b) forty percent, or (c) ninety-nine percent?

**❻** What is the world's most populous country: (a) the United States, (b) India, (c) the Soviet Union, or (d) China?

**❼** Which of the countries in the preceding question ranks second?

**❽** Match these European countries with their populations, as of 1990:

| Country | Population |
|---|---|
| 1. Austria | (a) 55.4 million |
| 2. Belgium | (b) 57.4 million |
| 3. France | (c) 7.5 million |
| 4. Italy | (d) 4.2 million |
| 5. Norway | (e) 9.9 million |

**9** The Pyrenees mountains in northern Spain are inhabited by a distinctive group of people, the Basques. What is unusual about their language?

**10** One of the least populated countries in the world, with only three people per square mile, lies next to the country with the largest population. What country is this?

**11** In terms of government, what do these countries have in common: Bhutan, Jordan, Saudi Arabia, and Spain?

**12** One European country has, not one, but *four* official languages. What is this country? For extra credit, what are the languages?

**13** Most of South America was colonized by Spaniards who explored the continent in the fifteenth and sixteenth centuries. Therefore, Spanish is the official language of most of the present countries of South America. Can you name the countries in which Spanish is *not* the official language?

**14** The largest country in Central Africa is Zaire, which was formerly known as the Belgian Congo. Among the two hundred groups of natives roaming its jungles is one group greatly respected for individual bravery in facing much larger opponents. Can you name and describe these people?

**⓯** Within the Arctic Circle, in a vast region that stretches across northern Norway, Sweden, and Finland, live the remnants of an ancient people. Only about 35,000 exist today, hunting small game and raising huge herds of reindeer. What is their land called?

**⓰** In all of South America and Central America, there is only one country that has a strong history of true democracy, where elections are peaceful and honest, and leaders are chosen by popular vote. What country is that? For extra credit, where is it located?

**⓱** In all of Africa, only one country, the Republic of Botswana, has been a democracy for more than twenty-five years. Where is Botswana?

**⓲** In South America, Paraguay held a national vote for the first time in 1989 and Chile elected its first democratic leader since 1973. For this two-parter: (a) where is Paraguay, and (b) where is Chile?

**⓳** In an unprecedented wave of change during 1989, Communist leaders were ousted in adjoining countries in east central Europe. Name the three countries in which this occurred.

**⓴** Most of the one hundred million people in North Africa are Arabic-speaking Muslims, giving them a spiritual tie to the people of the Middle East. What countries comprise this group?

**21** Nearly half of the people in this country just south of Mexico and northwest of Honduras and El Salvador are descendents of the Mayan Indians, whose intricate culture flowered until the people were subdued by Spanish conquerors. Name the country.

**22** The Sandinistas, a group originally founded to overthrow the corrupt regime of a dictator, have been much in the news in recent years. Although they succeeded in their goal, they later instigated a devastating civil war throughout the country. What was the country? For extra credit, where is it located?

**23** For thousands of years, the population of this huge kingdom in the Middle East was comprised of desert nomads called Bedouins, who moved from place to place. Now they live largely in cities such as Riyadh, the capital, Jidda, Mecca, and Dhahran. For this two-parter: (a) what is the country, and (b) where is it located?

**24** These cities in the developing world, commonly known as the Third World, are known, unfortunately, for the fact that they have large populations of people who live in poverty in their slums. Match the cities with the countries in which they are located:

| City | Country |
|------|---------|
| **1.** Rio de Janeiro | **(a)** Nigeria |
| **2.** Addis Ababa | **(b)** Iran |
| **3.** Dhaka | **(c)** Brazil |
| **4.** Lagos | **(d)** Bangladesh |
| **5.** Teheran | **(e)** Ethiopia |

**㉕** There were twice as many people living in the Republic of Ireland 150 years ago as there are today. People abandoned farms and left rural areas to migrate to other countries where they could enjoy a living wage. Many moved to cities, such as the capital, to try to escape poverty. What is the name of the capital?

**㉖** More than a billion people are crowded into one region that includes the countries of Afghanistan, Bangladesh, Bhutan, India, the Maldives, Nepal, Pakistan, and Sri Lanka. What is this geographic region called?

**㉗** The independent state with the smallest population in the world has another unique distinction: it has a zero birth rate. What is its name?

**㉘** This Portuguese province, located near the mouth of the Canton River on the southeast coast of China, is the most densely populated place on earth, with more than 72,000 people per square mile within its boundaries. Name it.

**29.** One country in North America has two official languages because of the linguistic background of two segments of the population. For this two-parter: (a) what is the country, and (b) what are the languages?

**30** The United Nations, headquartered in New York City, was formed to help bring the peoples of the world into greater harmony and advance the cause of peace. How many nations are officially members of the United Nations: (a) 89, (b) 112, (c) 138, or (d) 159?

**31** The least crowded country in South America is the Republic of Suriname (Surinam), which is slightly larger than the state of Georgia and has a population of only 400,000 people. Where is Suriname?

**32** The people who are native to the Andes of South America and live in the mountain plateaus have adapted physiologically, over the years, to the high altitudes. What form has this adaptation taken?

**33** The "Center of Population" is that point at which a country would balance if it were a rigid plane and the entire population was distributed upon it in the various locations where people live. In the United States is this center located: (a) in the middle of Missouri, (b) in the western part of Kansas, or (c) in southern Minnesota?

**34** The cities with the five largest populations in the United States are: Chicago, Houston, Los Angeles, New York, and Philadelphia. Can you rank them by numbers?

**35** In general terms, where do you find those people whom we refer to as "Latin Americans"?

**36** Ancient civilizations occupied most regions that are now known as specific countries, but which came into being long after those ancient civilizations flourished. Try to match the following examples of ancient civilizations with the countries that now occupy all or part of their territories:

| Civilization | Country |
|---|---|
| 1. Mayan | (a) Brazil |
| 2. Minoan | (b) Peru |
| 3. Incan | (c) United States |
| 4. Aztec | (d) Mexico |
| 5. Guarani | (e) Guatemala |
| 6. Anasazian | (f) Greece |

**37** Match the countries with the titles of the heads of their government.

| Country | Government Head |
|---|---|
| 1. Mongolia | (a) Governor General |
| 2. Mexico | (b) Emperor |
| 3. Japan | (c) Chairman |
| 4. The Bahamas | (d) President |
| 5. Austria | (e) Chancellor |

**38** The largest single racial group in Guyana is made up of East Indians, which is unusual because the East Indies lie halfway around the world. Where is Guyana? For extra credit, what was Guyana formerly called?

**39** Match the peoples listed below with the countries in which they dwell:

| People | Country |
|--------|---------|
| **1.** Iberians | **(a)** Yugoslavia |
| **2.** Mestizos | **(b)** Hungary |
| **3.** Serbs | **(c)** Spain |
| **4.** Maoris | **(d)** Venezuela |
| **5.** Magyars | **(e)** New Zealand |

**40** The people of Spain stem largely from the following groups, which migrated in the past from other regions. Try to match the names of the groups on the left with their origins:

| Group | Origin |
|-------|--------|
| **1.** Carthaginians | **(a)** Ireland and Wales |
| **2.** Celts | **(b)** Italy |
| **3.** Moors | **(c)** Germany |
| **4.** Romans | **(d)** Tunisia |
| **5.** Visigoths | **(e)** Mauritania |

**41** The island of Corsica in the Mediterranean west of Italy was settled by Ligurians, Etruscans, and Carthaginians, later formed part of the Roman Empire, and was also occupied at one time or another by people from Germany, France, and the Middle East. What country owns Corsica today?

**42** The Japanese constitute a remarkably homogeneous group, with fewer mixtures of people from other regions than almost any country in the world. Many Japanese had forebears who were Ainus, aborigines who inhabited the islands a long time ago, and who are now concentrated largely on the northernmost of the four major Japanese islands. Name this island.

**43** In an earlier chapter, Assyria was described as an ancient country that was mankind's first civilization around 5000 B.C., but that no longer exists. However, since the fall of that empire some two thousand years ago, Assyrians have lived in the land of their forefathers with Arabs, Kurds, Persians, and others. Where do the majority of Assyrians live today?

**44** One of the great, persistent problems in the Middle East relates to the Palestinians and their desire to establish themselves in the land they feel is rightfully theirs. Where is Palestine?

**45** Many European nations, becoming overcrowded and finding little room at home for economic expansion, either seized or purchased lands overseas in which to develop their colonies. Match the countries below with the colonies that belong, or belonged, to them:

| Country | Colony |
|---|---|
| **1.** France | **(a)** Falkland Islands |
| **2.** Great Britain | **(b)** Sardinia |
| **3.** Spain | **(c)** Svalbard |
| **4.** Italy | **(d)** Balearic Islands |
| **5.** Sweden | **(e)** Guadeloupe |

**46** Gibraltar has been a subject of controversy for centuries. For a three-parter: (a) where is it, (b) what country owns it, (c) what other country has been trying for many years to get control of it?

**47** This continent has been in constant political turbulence ever since the late nineteenth century when France, Germany, Portugal, Great Britain, Italy, and Turkey moved to colonize and partition its many nations and regions. Name the continent.

**48** The recent unification of East Germany and West Germany often brought the two countries into the news. To find out how much you really know about them, answer the following True/False questions:

(a) True or false. West Germany was larger than East Germany.

(b) True or false. The populations were about the same.

(c) True or false. Leipzig and Dresden were in East Berlin.

(d) True or false. The Capital of West Germany was West Berlin.

(e) True or false. The official title of East Germany was the German People's Republic.

(f) True or false. The Official title of West Germany was the German Federal Democracy.

(g) True or false. The republics of East Germany and West Germany were proclaimed in 1949.

**49** The geography and economics of the world in recent decades have been influenced greatly by increasing demands for energy and the production of petroleum. In this regard, the organization called OPEC is mentioned almost daily in the press. For this two-parter: (a) what do the OPEC initials stand for, and (b) what countries are members of OPEC? (Name at least three).

# Answers

① *(b) The United States population is currently estimated to be about 250,000,000.*

② *(b) At the end of the 1980s, statisticians estimated that the world's population was about 5,400,000,000.*

③ *(c) In 1950 the earth's total population was about 2,500,000,000. So, by 1990, the planet's population had more than doubled!*

④ *(b) 6,127,000,000.*

⑤ *(c) Nearly all Americans are descendents of immigrant ancestors. Only the direct descendents of Native Americans, less than one percent of the population of the United States, can claim that their ancestors are "native" to America.*

⑥ *(d) About one-fifth of the earth's people live in China.*

⑦ *India. It is growing at such a rate that its population may soon surpass that of China.*

⑧ *1(c) Austria/7.5 million, 2(e) Belgium/9.9. million, 3(a) France/55.4 million, 4(b) Italy/57.4 million, and 5(d) Norway/ 4.2 million.*

⑨ *The Basque language has eight dialects and is not related to any other existing language.*

⑩ *Mongolia, which averages only three people per square mile.*

(11) *They are monarchies.*

(12) *Switzerland. The languages are French, German, Italian, and Romansh (a dialect derived from Roman times).*

(13) *The exceptions are Brazil, where Portuguese is spoken, Guyana, which is largely English-speaking, Suriname, which is Dutch, and French Guiana, where French is the major language.*

(14) *These are the Pygmies, sometimes called Negrillos, who are of slight build and generally less than five feet in height.*

(15) *Lapland.*

(16) *Costa Rica. It is located just above Panama, with Nicaragua to the north, and bordered by the Pacific Ocean on the west and the Caribbean Sea on the east.*

(17) *Botswana, slightly larger than Texas, is in southern Africa, inland, just north of the Republic of South Africa.*

(18) *(a) Paraguay lies high in the Andes of South America, completely landlocked, with Bolivia to the north, Argentina to the south, and Brazil to the east. (b) Chile occupies a large portion of the western coast of South America, from the Peruvian border to the southern tip of the continent.*

⑲ *Communist leaders were ousted in Poland, Czechoslovakia, and Romania.*

⑳ *Egypt, Libya, Algeria, Tunisia, and Morocco.*

㉑ *Guatemala, which lies on the Pacific Ocean and also has a small coastline on the Caribbean Sea.*

㉒ *Nicaragua. It lies between Costa Rica to the south and Honduras to the north and has shorelines on both the Pacific and the Caribbean.*

㉓ *(a) Saudi Arabia. (b) It occupies most of the Arabian Peninsula, with the Red Sea to the west, the Persian Gulf to the east, and the Arabian Sea to the south.*

㉔ *1(c) Rio de Janeiro/ Brazil, 2(e) Addis Ababa/ Ethiopia, 3(d) Dhaka/ Bangladesh, 4(a) Lagos/Nigeria, and 5(b) Teheran/Iran.*

㉕ *Dublin.*

㉖ *South Asia.*

㉗ *The Vatican. It has a population of 750, largely composed of religious members of the Roman Catholic Church.*

㉘ *Macao, which consists of a peninsula and two small islands with a total area of six square miles, has an estimated population of 433,000.*

㉙ *(a) The country is Canada. (b) The languages are English and French.*

㉚ *(d) The count, at the beginning of the 1990s, was 159.*

㉛ *Suriname lies on the Caribbean Sea in the northeastern region of South America.*

㉜ *To adapt to the thin air found at altitudes from about 12,000 feet to as high as 18,000 feet, these people have developed lungs that are larger than normal.*

㉝ *(a) The Center of Population, which shifts slightly from one decade to the next, is now located in the middle of Missouri.*

㉞ *New York, Los Angeles, Chicago, Houston, and Philadelphia.*

㉟ *In Spanish- and Portuguese-speaking countries south of the United States, in the West Indies, Central America, and South America.*

㊱ *1(e) Mayan/Guatemala, 2(f) Minoan/Greece, 3(b) Incan/Peru, 4(d) Aztec/Mexico, 5(a) Guarani/Brazil, and 6(c) Anasazian/United States.*

㊲ *1(c) Mongolia/Chairman, 2(d) Mexico/President, 3(b) Japan/Emperor, 4(a) The Bahamas/Governor General, and 5(e) Austria/Chancellor.*

(38) *Guyana is located on the north coast of South America. The East Indians were imported there by the British as indentured servants in the early part of the nineteenth century. It was formerly called British Guiana.*

(39) *1(c) Iberians/Spain, 2(d) Mestizos/Venezuela, 3(a) Serbs/Yugoslavia, 4(e) Maoris/New Zealand, and 5(b) Magyars/Hungary.*

(40) *1(d) Carthaginians/Tunisia, 2(a) Celts/Ireland and Wales, 3(e) Moors/Mauretania, 4(b) Romans/Italy, and 5(c) Visigoths/Germany.*

(41) *France. The island is officially a region of France comprising two departments.*

(42) *Hokkaido.*

(43) *Of the estimated three million Assyrians, more than half live in Iraq.*

(44) *Historic Palestine, originally known as the Holy Land, lies on the eastern shore of the Mediterranean Sea, comprising parts of modern Egypt, Jordan, and Israel.*

(45) *1(e) France/Guadeloupe, 2(a) Great Britain/Falkland Islands, 3(d) Spain/Balearic Islands, 4(b) Italy/Sardinia, and 5(c) Sweden/Svalbard.*

**46** *(a) Gibraltar is located on the north shore of the western entrance to the Mediterranean. (b) It is owned by Great Britain. (c) Spain has long been trying to acquire it.*

**47** *Africa.*

**48** *(a) True. West Germany occupied more than 95,000 square miles, and East Germany just under 42,000. (b) False. West Germany had more than 60 million people and East Germany only about 17 million. (c) True. They are old German cities, each with a population of more than half a million people. (d) False. The capital was Bonn. (e) False. The title was the German Democratic Republic. (f) False. The title was the Federal Republic of Germany. (g) True. West Germany was proclaimed on May 23, 1949 and East Germany on October 7, 1949.*

**49** *(a) OPEC stands for Organization of Petroleum Exporting Countries. (b) The members are Algeria, Ecuador, Gabon, Indonesia, Iran, Iraq, Kuwait, Libya, Nigeria, Qatar, Saudi Arabia, United Arab Emirates, and Venezuela.*

# Natural Disasters

Natural disasters are an essential part of geography for two fundamental reasons: first, because many land and water features can cause or contribute to the catastrophic events that take place and, second, because the disasters change the face of the globe, sometimes dramatically and decisively.

Among the "disaster" subjects included in this chapter are earthquakes, volcanoes, avalanches, tidal waves, floods, and catastrophic storms that can severely affect the terrain locally and regionally during their passage.

Let's start with volcanoes. How much do you know about these horrendous forces of nature?

**❶** The most devastating volcanic explosion in recorded history occurred on August 27, 1883, on an island in Asia. More than 36,000 people were killed and the noise of the blast was heard 3,000 miles away. (a) Name the island, and (b) describe its location.

**❷** Another historic volcanic explosion occurred in 1902 when Mount Pelee erupted, wiping out the city of St. Pierre at its foot and killing some 30,000 people. On what island did this disaster occur?

**❸** Match these volcanoes in Latin America with the countries in which they are located:

| Volcano | Country |
|---|---|
| **1.** Cotopaxi | **(a)** Chile |
| **2.** Nevado del Ruiz | **(b)** Ecuador |
| **3.** Poas | **(c)** Peru |
| **4.** Lascar | **(d)** Costa Rica |
| **5.** El Misti | **(e)** Colombia |

**❹** The greatest volcanic eruption in the United States in recorded history took place in May, 1980, when a blast that was described as "500 times greater than the bomb that fell on Hiroshima" destroyed one whole side of Mount Helena. Where is Mount Helena?

**❺** One of the greatest eruptions in ancient times (and possibly the most devastating that ever took place) occurred in 1470

B.C. on the island of Santorini, lying between Greece and Turkey. It wiped out a highly developed Minoan civilization that had spread to the island from nearby Crete, to its south. In what sea did this occur?

**6** Probably the best known eruption in history was the one that occurred on August 24, 79 A.D. It destroyed Pompeii, Herculaneum, and several villages. Not only was it described in great detail by writers of the day, but the ashes entombed and preserved the bodies of many of the victims, whose remains can be seen to this day. For this three-parter: (a) what is the name of the volcano, (b) in what country did this occur, and (c) where is the volcano located?

**7** How many active volcanoes are there in North America: (a) 10, (b) 20, (c) 30, or (d) 50?

**8** More than seventy-five percent of the active volcanoes in the world lie within a zone running along the west coast of the Americas from Chile to Alaska and down the east coast of Asia from Siberia to New Zealand. What is this zone called?

**9** About 7,000 years ago, Mazama, a 9,900-foot-high volcano in Oregon erupted violently, causing the top of the mountain to collapse and leaving a hole six miles across and half a mile deep. This eventually filled with rain water to form a lake that is now a popular tourist attraction and a national park. Can you name this lake?

**10** One of the most active volcanoes in the United States is Mauna Loa, which rises 13,680 feet and has two of the world's largest craters, Kilauea and Mokuaweoweo. Mauna Loa is in Hawaii, but on which island?

**11** One of the most copious products of volcanic action is called magma. Is magma: (a) molten lava that flows from the volcano, (b) clouds of deadly gases that can suffocate anyone who ventures too close, or (c) huge puffs of soot and ash that blanket the surrounding countryside?

**12** Lava often forms in great, columnar sheets of a rocky substance. One of the most notable examples is seen in the Palisades in New Jersey, which look like the ramparts of a mighty fortress along the Hudson River. What is this rocky substance called?

**13** The largest crater in the world is said to be the one on 5,200-foot Mount Aso, located on a Japanese island just southeast of Korea. Name the island.

**14** In the United States, one of the most dramatic examples of lava formations is in the Black Hills of northeastern Wyoming. Dwarfing the forest at its base, it is a startling monolith of clustered lava columns rising 865 feet into the sky. What is its name?

**Now let's explore some facts about another of nature's most powerful forces: earthquakes.**

**⑮** The earthquake that caused the largest death toll in history was one that killed some 830,000 people in 1556. In what country did it occur?

**⑯** In the matter of casualties, the worst earthquake in the United States was a "drop in the bucket" compared with major quakes in other countries. But it caused millions of dollars worth of damage. This, the most infamous of American earthquakes struck a little after five on the morning of April 18, 1906. Where did it strike?

**⑰** One of the factors associated with the cause of earthquakes is called a fault. Is a fault: (a) a break in a stratum of rock or a surface along which rock bodies have been displaced, (b) a zone of pressure caused by subterranean upheavals that cause the earth to buckle, or (c) sections underground where hard rock has become decayed and porous?

**⑱** One of the major faults in the United States is located in California, extending from San Francisco south to the Colorado Desert. What is its name?

**19** Since the magnitude of earthquakes generally determines the amount of damage, or potential damage, that can be caused, each one is measured by instruments and reported in a comparative manner. For this two-parter: (a) what is the name of the most common form of measurement used, and (b) how is the intensity reported, comparatively?

**20** What is the name used for a scientist who specializes in the study of earthquakes: (a) geologist, (b) speliologist, (c) meteorologist, or (d) seismologist?

**21** When a major earthquake occurs, it is invariably followed by a series of additional earthquakes that are generally of lesser magnitude but may still cause considerable damage. What are these called?

**22** The most devastating earthquake in the past decade struck a region of the Soviet Union on December 7, 1988. It killed more than 25,000 people and left some 500,000 homeless. For this two-parter: (a) what is the name for this region of the USSR, and (b) exactly where is it located?

**23** Earthquakes are infrequent on the Atlantic Coast of the United States and generally much less severe than those in the western part of the country. The worst quake ever to hit the East occurred on August 31, 1886, in South Carolina, heavily damaging one of its major cities. Name the city.

**24** As is the case of volcanic action (see question 8), earthquakes are found much more frequently in certain zones of the world than in others. One of these, in southeastern Asia, runs in part along the Indonesian archipelago. What is this zone called?

**25** The incredible power of an earthquake was seen in one that occurred in the western United States in 1959. The shock jolted some 500,000 square miles of largely wilderness area, tearing off enormous chunks of 8,257-foot Mount Jackson in Yellowstone Park, creating a new lake, and reactivating 160 dormant geysers. In what states did this take place?

**26** Even in ancient times, people worried about quakes when horses began rearing unexpectedly, fish in pools became strangely agitated, or normally docile pigs tried to jump out of their pens. What are some of the other short-term precursors that signal imminent earthquakes?

**27** Although erupting volcanoes very often trigger earthquakes, infrequently the process will be reversed and an earthquake will cause a sleeping volcano to become active. Such was the case in November 1975, when a fault ruptured in Hawaii some twenty-five miles south of Hilo. What was the name of the volcano?

**28** In September 1923 Japan was devastated by one of history's greatest quakes. It registered 8.3 on the Richter Scale, caused some 200,000 deaths, and shattered more than 5,000 buildings in Tokyo alone. One of the few buildings that survived was the newly completed Imperial Hotel, which suffered no damage at all. It had been designed as an earthquake-proof structure by one of America's greatest architects, who had studied earthquake problems and solutions in great detail. Can you name him?

**29** Volcanoes and earthquakes, jointly and separately, often produce another natural cataclysm that can be deadlier than either of them. It is called a *tsunami*. Is a tsunami: (a) a dense cloud of gas that issues from cracks and fissures in the earth and rolls across the land, suffocating all living things in its path, (b) a windstorm of cyclonic force created by heat or pressure, or both, that sweeps across the terrain almost without warning, or (c) a giant, high-velocity sea wave generated by an earthquake or volcanic eruption, which can travel great distances and cause havoc when it reaches any shorelines in its path?

**30** In 1946 an earthquake at Unimak Island off the northwest coast of the United States produced a tidal wave that traveled all the way to the Hawaiian Islands where, without advance warning, it killed 159 people and caused millions of dollars of damage. For this two-parter: (a) where is Unimak Island and, (b) about how far did the wave have to travel to cause casualties and destruction?

**31** If you were in a small boat in the open ocean and an earthquake occurred in the sea floor not far away, what effects would you feel?

**32** In November 1755 the city of Lisbon suffered a double disaster when an earthquake struck, destroying five of every six buildings and triggering an enormous tidal wave along the nearby Tagus River. In both disasters, a total of 60,000 people perished. In what region of what country is Lisbon located?

**Now let's talk about floods and hurricanes.**

**33** When it comes to water-borne Doomsdays, nothing can eclipse some of the great, historic floods of the world. The worst on record in modern times occurred in 1931 when massive floods swept along the Yellow River and drowned more than three million people, most of them peasants in small villages who had no possible means of escape. (a) Where is the Yellow River, and (b) what is its Chinese name?

**34** The most infamous flood in American history took place on May 31, 1889, when a reservoir burst spilling millions of tons of water into the Conemaugh River in southwestern Pennsylvania. In the ensuing flood, 2,200 people were drowned. What was the city that was hardest hit?

**35** About twice that number were killed eleven years later in a flood in southeastern Texas. However, many of these victims were killed as a result of the tropical hurricane that caused the flood to begin with and resulted in tremendous property damage. Name the city that was devastated. For extra credit, tell why this city was so vulnerable to both floods and hurricanes.

**36** What river in North America has been most noted, historically and right up to the present, for its tendency to flood?

**37** In 1948 one of the largest rivers in North America, the Columbia, went on a rampage. It wiped out entire towns as it (and its tributaries) flooded, and left some 48,000 people homeless. In what two states did most of this catastrophe take place?

**38** Along broad rivers that are bordered by flatlands vulnerable to floods, the inhabitants have devised a relatively simple and inexpensive impediment to flooding. What is it called?

**39** Geologists, geographers, government administrators, and insurance adjusters categorize major rivers and river systems by speaking of "two-year floods," "ten-year floods," and "fifty-year floods." Which of these types of floods is the most severe? For extra credit, why?

**40** In 1966 the Arno River in Italy overflowed its banks and caused great damage to one of the finest Renaissance cities in all of Europe. Many priceless landmarks were heavily damaged. Name the city.

**41** Another notable Italian city has become increasingly subject to flooding, but not from a river. Name (a) the city, and (b) the body of water that causes the floods.

**42** The following places in the United States have also been devastated by major floods. Match the places with the states in which they are located:

| Place | State |
|---|---|
| **1.** Lake Okeechobee | **(a)** California |
| **2.** Buffalo Creek | **(b)** South Dakota |
| **3.** Putnam | **(c)** Florida |
| **4.** Saugus | **(d)** Connecticut |
| **5.** Rapid City | **(e)** West Virginia |

**43** What are flash floods and why are they so vicious?

**44** Whenever a natural disaster of major proportions occurs, the region affected is likely to come under the scrutiny of local geomorphologists. Is a geomorphologist: (a) a specialist who studies the origins, development, and changes that take place in landforms, (b) an historian who keeps records of casualties and property damage in places hit by such catastrophes, or (c) a geographer, generally with an engineering background, who helps to plan ways to restore areas that have been ravaged by nature?

**45** When it comes to windstorms, the greatest cataclysm within the past two centuries occurred on November 13, 1970, when a cyclone resulted in the deaths of some 300,000 victims in southern Asia, on the north bend of the Bay of Bengal. Name the country in which this catastrophe occurred.

**46** The most disastrous hurricane to hit solely the continental United States was one that occurred on September 21, 1938. Besides inflicting immense amounts of property damage, it caused some 600 deaths, largely because the storm occurred in an area that historically had been hurricane-free and was, therefore, not anticipated. Where did this hurricane hit?

**47** Hurricanes and typhoons generally move westward at about ten miles per hour and then curve poleward as they approach the western boundaries of the oceans at 10° to 30° latitude. They can last as little as a day or two before weakening into an ordinary storm as they pass over land. But for how long can these storms continue at hurricane force before dying out: (a) seven days, (b) ten days, (c) eighteen days, or (d) twenty-eight days?

**48** The most costly hurricane in American history was Hurricane *Agnes*, which continued wreaking havoc for eleven days in late June 1972. Although advance warnings kept the death toll to 118, property damage mounted to more than three billion dollars. Where did *Agnes* strike?

**49** Away from the coastal areas of the United States, one type of very deadly storm killed 350 people in Alabama, Georgia, Kentucky, Ohio and Tennessee on May 27, 1973, and 689 people in Missouri, Indiana, and Illinois on March 18, 1925. What kind of windstorm was this?

**50** From August 30 to September 7, 1979, one of the worst hurricanes ever to hit North America ravaged the Caribbean and the East Coast of the United States, causing damage in the billions and the loss of at least 1,100 lives. What was the name of this hurricane?

**51** In 1906 a typhoon struck Hong Kong, killing some 10,000 people. What is a typhoon?

**52** The worst snowstorm on record in the United States was the Blizzard of '88, which struck the eastern part of the country in mid-March. How is a blizzard different from an ordinary snowstorm?

**53** On May 25, 1955, a sudden storm skipped across Kansas, Missouri, Oklahoma, and Texas, leaving 115 people dead in its wake. The storm was typical of ones that occur regularly in these regions. What was it?

**54** The topography and terrain of mountain regions are constantly affected by avalanches and landslides which, though dramatic, are seldom accompanied by any great loss of life or property damage. A notable exception was an avalanche on the slopes of 22,200-foot Mount Huascarin in Peru on January 10, 1962. It buried the entire village of Ranrahirca and smothered some 4,000 victims. In what mountain range did this take place?

**55** Our last question relates to a dire trend that alarms scientists and laypersons alike. It is the process whereby excessive heat is being trapped in the atmosphere by pollutant gas emissions. Besides being harmful to flora and fauna, this undesirable trend makes many areas of the world vulnerable to rising sea levels that could change the whole face of geography. What is the name of this much-feared phenomenon?

## Answers

① *(a) Krakatao, (b) in the Indian Ocean.*

② *This volcanic explosion occurred on Martinique, in the West Indies.*

③ *1(b) Cotopaxi/Ecuador, 2(e) Nevado del Ruiz/Colombia, 3(d) Poas/Costa Rica, 4(a) Lascar/Chile, and 5(c) El Misti/Peru.*

④ *In the Cascade Range in the state of Washington, about fifty miles northeast of Portland.*

⑤ *The Aegean Sea.*

⑥ *(a) Vesuvius, (b) Italy, and (c) near Naples, on the shores of the Bay of Naples.*

⑦ *(b) There are twenty active volcanoes, all of them along the Pacific Coast.*

⑧ *The "Ring of Fire."*

⑨ *Crater Lake, situated in Crater Lake National Park. The second deepest* lake in the United States, it is filled solely by rain and snowfalls.

⑩ *Mauna Loa is on ·the "Big Island," Hawaii.*

⑪ *(a) Magma is molten lava, which erupts explosively and is impregnated with gas.*

⑫ *It is basalt.*

⑬ *Kyushu, the southernmost of the Japanese islands.*

⑭ *Devil's Tower, in Devil's Tower National Monument.*

⑮ *In China (Shaanx).*

⑯ *In San Francisco.*

⑰ *(a) A fault is a break.*

⑱ *The San Andreas Fault, probably the most well-known fault in the United States.*

⑲ *(a) The Richter Scale, named after the American who devised it, Charles F. Richter. (b) The intensity is reported in degrees from one to ten, with ten being the most intense.*

⑳ *(d) Seismologist.*

㉑ *Aftershocks.*

㉒ *(a) Armenia. (b) It lies in southern Russia, near the Caucasus Mountains, bordered on the south by Turkey and Iran.*

(23) *Charleston, located on the coast of South Carolina.*

(24) *The Indo-Australian Plate, which lies along the equator and on the Indian Ocean.*

(25) *In southern Montana and northwestern Wyoming.*

(26) *In addition to the bizarre behavior of domestic animals, other signals that warn of an earthquake are an increase in hydrogen in the soil, flashes of bright lights in the sky (possibly related to the release of gases in the earth), and unexplained rises or drops in well water.*

(27) *The volcano was Mount Kilauea, which rises to 3,646 feet.*

(28) *Frank Lloyd Wright.*

(29) *(c)* Tsunami *is Japanese for "overflowing wave," and is also called a tidal wave or seismic sea wave. Such waves can travel 500 miles per hour and rise to heights of more than 250 feet when they hit the shore.*

(30) *(a) Unimak Island, which belongs to Alaska, is in the Aleutian Islands. (b) About 2,300 miles.*

(31) *Probably nothing. Even if a* tsunami *followed the eruption, the waves are long and rolling and hardly felt in the open sea.*

㉜ *Lisbon, the capital of Portugal, is situated in the southern part of the country and inland about seven miles from the Atlantic Ocean.*

㉝ *(a) The Yellow River is in central China, (b) The Huang Ho.*

㉞ *Johnstown, Pennsylvania, was the hardest hit city.*

㉟ *Galveston, in southeastern Texas. It is vulnerable because it sits on a long, narrow island that rises only a few feet above sea level in the Gulf of Mexico.*

㊱ *The Mississippi, which not only meanders through many lowlands, but is subject to enormous surges of water from heavy rainstorms and sudden spring thaws. These come both from the Mississippi's upstream course and from the many tributaries that feed into it.*

㊲ *Washington and Oregon, particularly along the border of the two states.*

㊳ *The levee. It is simply an embankment made of readily available earth and sometimes reinforced with native rocks and stumps.*

㊴ *Fifty-year floods are the greatest in severity. The experts reason that a river will have moderate flooding every couple of years, a sizeable flood at least once over the course of ten years, and probably a devasting flood at some time within the next half century.*

**40** *Florence.*

**41** *(a) Venice, (b) Gulf of Venice, on the Adriatic Sea.*

**42** *1(c) Lake Okeechobee/ Florida, 2(e) Buffalo Creek/West Virginia, 3(d) Putnam/Connecticut, 4(a) Saugus/California, and 5(b) Rapid City/South Dakota.*

**43** *Flash floods occur when there is a deluge of rain that fills gullies and small valleys and rushes down them. They are dangerous because they occur so quickly and without warning.*

**44** *(a) From* geo, *"earth,"* and morph, *"form."*

**45** *Bangladesh. This small country, roughly the size of Wisconsin, is surrounded on the north, west, and east by India.*

**46** *Along the east coast, most forcefully from Long Island, New York, north through the New England states.*

**47** *(d) It is not uncommon for a hurricane to continue in force for two or three weeks. Some have persisted for four weeks and a few for as much as thirty days.*

**48** Agnes *rampaged from Florida to New York.*

⁴⁹. *A tornado. Its deadliness lies in the fact that its funnel shape is small (less than a mile in diameter), but twisting furiously with winds that can reach 300 miles per hour.*

⁵⁰ David.

⁵¹ *A typhoon is the name given to a hurricane or tropical cyclone in the western Pacific and China seas.*

⁵² *A blizzard is a heavy snowstorm that occurs when the temperature is below 20 °F and there are high, gusty winds, more than thirty-five miles per hour.*

⁵³ *A tornado. It is most common in regions where there are prairies and other wide-open spaces across which the wind can whirl unobstructedly.*

⁵⁴ *The Andes.*

⁵⁵ *Global warming.*